# DAS TIERREICH

## Eine Zusammenstellung und Kennzeichnung der rezenten Tierformen

The ,,TIERREICH" was initiated by the German Zoological Society in order to create a complete systematical treatment of all groups of the recent animal kingdom. After the World War II this undertaking was stopped because it seemed impossible to attain its completeness. In consequence of that, however, also ceased both incitement and possibility of publishing comprehensive revisions of large animal groups in a continued series. Therefore the Verlag Walter de Gruyter & Cie. (Berlin) came to the conclusion to revive the ,,TIERREICH". The original views will be maintained, the work, however, is planned now as an unconstrained sequence of systematical monographs without intending to treat all zoological categories according to a fixed plan. These monographs should not be restricted only to the belongs of taxionomy but also have regard to anatomical, biological, physiological, and other facts as far as they are able to characterize the group in question without breaking the frame of the whole work. All authors who are willing to publish systematical revisions of recent animal groups are invited to collaborate. According to the wishes of the authors, their contributions will be printed in English, French, or German. The authors will get an honorarium of DM 55,— for each 16 printed pages and 5 free copies after come out of their contributions. Further details may be gathered from the exemplaries no. 72 and subsequents published since 1958; by request the publishers will give any informations or send exact directions for the authors. Summary revisions of zoological groups (checklists) will although be published in the series ,,DAS TIERREICH".

---

Le ,,TIERREICH" fut fondé pa la Société Allemande de Zoologie dans le but de réaliser une présentation complète de la systématique de tous les groupes zoologiques récents. Cette entreprise fut suspendue après la deuxième guerre mondiale car la présentation complète que l'on s'était fixée pour but semblait impossible à réaliser. Mais, de ce fait, disparut, avec la possibilité, le stimulant nécessaire à la publication continue d'amples révisions de groupes zoologiques importants. C'est pourquoi, les éditions Walter de Gruyter & Co. (Berlin) ont pris la décision de faire renaître le ,,TIERREICH" sous l'optique originale pour ce qui est du fond, certes, mais uniquement sous la forme d'une série de monographies systématiques, sans lien étroit entre elles dont le but ne sera pas de traiter tous les groupes zoologiques d'après un plan préconçu. Ces monographies ne doivent pas simplement se borner au domaine de la taxionomie, mais tenir compte également des faits anatomiques, biologiques, physiologiques et autres dans la mesure où ils sont susceptibles de mieux caractériser le groupe

# DAS TIERREICH

**Eine Zusammenstellung und Kennzeichnung der rezenten Tierformen**

Herausgeber

Robert Mertens
(Wirbeltiere)

Willi Hennig
(Wirbellose)

Schriftleitung
Heinz Wermuth

Unter Mitarbeit zahlreicher Spezialisten

„Πάντα ῥεῖ" „Sine systemate chaos"

**Lieferung 97**

**Lepidoptera Noctuiformes**

## Agaristidae I

(Palaearctic and Oriental Genera)

Bearbeitet von

**Prof. Dr. S. G. Kiriakoff, Gent**

Mit 39 Figuren im Text

**Walter de Gruyter · Berlin · New York**
**1977**

| Das Tierreich | Berlin | Lieferung 97 | Seite I—IX, 1—180 | 1977 |
|---|---|---|---|---|

ISBN 3 11 005876 6


Printed in Germany.
Satz: IBM-Composer Walter de Gruyter & Co., Berlin.
Druck: Karl Gerike, Berlin.
Bindearbeiten: Dieter Mikolai, Berlin.

# Familia Agaristidae I

## (Palaearctic and Oriental Genera)

Bearbeitet von

**Prof. Dr. S. G. Kiriakoff**

## Systematic Index

Fam. **Agaristidae**

I. Palaearctic and Oriental Genera

# Introduction

The structure of the antennae has been generally considered as the main character differentiating the Agaristid moths from the bulk of the Noctuids. HAMPSON in his Catalogue, part 3 and supplement 2 has followed this usage and placed a number of genera dealt with in the present work in the family Noctuidae, subfamily Acronyctinae (Catalogue part 9). JORDAN, and later DRAUDT (both in SEITZ's Macrolepidoptera of the World) have not followed HAMPSON: their Agaristidae (of Agaristinae) include genera with the typical "agaristine" antennae along genera with filiform, ciliate and pectinate antennae. STRAND in Lepidopterorum Catalogus, part 5, was still more comprehensive, and put in his subfamily Agaristinae, besides the above, quite a number of genera placed by every other systematist with the true Noctuidae.

The author of this revision, having examined the tympanic organs of most of the Agaristid genera, found that these structures possess in all the genera studied a Bulla abdominalis, an apomorphic structure not found elsewhere. This synapomorphy, quite independant of the antennal structure, was considered by the writer as a reliable criterion for placing the genera involved in a group taxonomically equal to that of the true Noctuidae. Both groups might be considered as subfamilies or families, the designation not mattering in so far as they are placed on the same taxonomic niveau. Indeed, we have to deal with sister-groups, of which the Agaristids are apomorphic, the Noctuids plesiomorphic. The complex belongs to the taxon Noctuiformes, the composition of which has been dealt with elsewhere, and need not to be described here.

The origin of the Agaristids should be sought somewhere in the eastern Palaearctic region of to-day, probably in an early period of the Caenozoic, long before the differentiation in a Palaearctic and an Indo-Australian or Oriental faunistic region. We have to admit a primary palaeotropical expansion, with a radiation in the direction of the New World. There still is an affinity between the chinese *Syfania* and the nearctic complex of *Alypia*, the latter probably being the ancestral group of the American Agaristid genera. A somewhat later radiation must have occured westwards from the cradle-region, producing the Ethiopian and the Madagascan Agaristid faunas. The Agaristidae not being a group of Gondwanian origin, the South African genera (as well as the Neotropical and the Australo-Oceanian ones) should be considered as the more recent.

The existence of several distinct Agaristid faunas (genera common to two or more faunas are but very few, and should be considered as belonging to an older, or more plesiomorphic branch of each fauna) allows us to propose a peculiar kind of taxonomic arrangement, not based on morphological but on spatial or chorological criteria. The writer had formerly used the space-criterion for two other groups of moths (the Notodontidae and the Thaumetopoeidae) and feels justified in proposing an arrangement of the same kind for the Agaristidae. It should be understood that the Agaristidae being morphologically a very homogenous group, no reliable morphological features can be used for the separation of the faunistic groups. Differences on a generic niveau can of course be noted, and there is, on the other hand, a parallel development in some regions towards sound-producing forms. The necessary details will be given in the introduction to each of the three main faunas. Generally speaking, however, it was deemed preferable not to attempt a well-nigh impossible arrangement on current cladistic lines, i. e. a succession of sister-groups, one plesiomorphic, the other apomorphic, and to resort to a differentiation on purely zoogeographical grounds, as stated above. As the writer had repeatedly pointed out in various publications, it is his conviction that the space or distributional factor should be considered as a differentiating or taxonomic character even, and in many instances more significant that the holomorphic characters generally used.

Of the groups recognised here, the Ethiopian (with its subdivision the Madagascan) fauna does possess the highest percentage of autochtonous apomorphic genera. The Oriental fauna has developed but a small number of apomorphic genera, including *Episteme* and a few sound-producing forms, besides a number of surprisingly homogenous insular genera. Unlike in the family Notodontidae, there is not a distinct, apomorphic Australian faunistic group, and Australian genera hardly differ from those of the continental and insular East Indies. Although seemingly older than the peripheric Old World groups, the American fauna shows very little apomorphies. The Nearctic genera are few and very much akin the Palaearctic ones; the somewhat aberrant genus generally known as *Euthisanotia* has proved to be heterogenous; and the Nearctic genera have remained surprisingly conservative. Being based on zoogeographical criteria, the groups recognized here, are of course taxa of a nature differing from that of the units generally used in a classification, which all are of a morphological nature, those used in a conventional-phylogenetic as well in a cladistic arrangement, not to speak of the OTUs used by the pheneticists. The taxonomic units used here are accordingly being termed Chorotaxa as distinguishable from the usual Morphotaxa. The writer does not know whether the chorotaxa-concept can be applied to all groups of animals, but it does seem to him that to try and find that out is worth while. In this revision the three following Chorotaxa are recognized, viz.:

1. Agaristinae: Palaearctic-Oriental
2. Heracliinae: Ethiopian-Madagascan
3. Alypiinae: Nearctic-Neotropic

Particulars about each group will be given in the respective parts of the revision.

Sincere acknowledgments are due to D. S. FLETCHER, P. VIETTE, A. DIAKONOFF, H. J. HANNEMANN, P. BASILEWSKY, and H. CLENCH who each have helped the writer in various, but uniformly efficient ways. Particular thanks go to H. INOUE, who kindly sent a pair of *Maikona jezoensis*. Mrs. M. LANE has dissected and drawn most of the specimens figured; a few were prepared by P. VIETTE; a couple by the writer; and one by Harry CLENCH.

H. WERMUTH, editor-in-chief of "DAS TIERREICH" is to be thanked most heartily for his constant help. W. HENNIG, editor of the Invertebrata, has kindly pointed out some imperfections in the presentation. The publishers have as usually attended to a perfect presentation of the volume. Last but not least, a subsidy from the Belgian National Fund for scientific research has made possible the preparation of the book, for which the writer is very grateful.

LEPIDOPTERA NOCTUIFORMES

Superfamilia NOCTUOIDEA

## Familia **Agaristidae**

1833 Agaristidae BOISDUVAL, (Agaristides) Faun. Entomol. Madagascar: 70; Ann. Mus. Paris, 2:218.
1840 Agaristites BLANCHARD, (Agaristes in the next column), Hist. Nat. Ins., 3:471.
1912 Agaristinae STRAND in Lepidopt. Cat., 5:3.
1920 Phalaenoididae HAMPSON, Cat. Lepidopt. Phal. Brit. Mus., Suppl. 2:534.

Diagnosis: Proboscis well developed; palpi mostly well developed, upturned, the third joint usually naked and porrect; frons with a rounded or conical prominence or process; antennae typically filiform and more or less distinctly dilated distally, before the extremity, but in a number of genera not dilated, ciliated or even pectinate; ocelli present; eyes sometimes hairy. Tympanic organs present, of the noctuoid type, but apomorphic in the presence, at the base of the abdomen, of a more or less well developed paired vesicle (Bulla abdominalis). Tibiae rarely spined; hind tibiae with two pair generally well developed spurs. Wings relatively broad, with the fore wings rarely produced apically, and bearing sometimes in male a sound producing apparatus of ribbed membrane. Nervation: in the fore wings, first and second anals (1 and 1 a) separate, third anal absent; vein 5 from or close to angle of cell; an areole present in nearly all genera. Hind wings with the first anal present, the third anal absent, vein 5 generally obsolescent, from the middle or near the middle of the discocellulars; veins 6, 7 from the upper angle of cell, rarely stalked; vein 8 free at the base, then bent downwards to anastomose with the cell at a point only, very rarely to beyond middle (*Pseudospiris*).

For most of the species, the early stages are unknown. The known larvae have all the prolegs present; an enlarged eleventh segment; tubercles as in the family Noctuidae, low, conoidal, smooth, each bearing a single hair. The pupa naked.

The following species described as Agaristidae have not been included in part one of the present monograph:

*Cnissocnema neuhaussi* BRYK, Arch. Naturgesch., **79** A(6): 124–126, 1913.
*Depalpata mirabilis* ROTHSCHILD, Ann. Mag. Nat. Hist. (9) **3**: 488, 1919. Both belong to the Geometridae.
*Paraegocera variata* CANDEZE, Encycl. Entomol., B. IV, Lépidopt., 2: 79 (1927).

The single male described and said to be in the collection of the University of Liège could not be located there and should be considered as lost. It most certainly does not belong to the Ethiopian genus *Paraegocera* HAMPSON, and his taxonomic position cannot be determined with certainty.

## I. Palaearctic and Oriental Genera

The rather homogenous group of Palaearctic-Oriental Agaristidae can be divided into two complexes, the one containing the single genus *Episteme* HÜBNER, the other the remaining genera. *Episteme* is distinguished by several characters. Its species have a large, well-developed, rather elaborate, roughly V-shaped sternal plate of the VIIIth urite, while the other Agaristid genera have that structure rather poorly differentiated; the uncus is relatively very small and lanceshaped, not approximated by the corresponding structure in the remaining genera; a third distinctive feature is the vertical position of the harpe, mostly extending beyond the costal margin of the valva, and not placed in the axis of the latter as in the other Agaristidae.

It is not yet possible to say whether both groups are to be placed on the same taxonomic level. *Episteme* is possibly a younger branch that has split off somewhere in South-east Asia, and is undoubtedly monophyletic. As we shall see, the other group is composed of a number of subgroups some of which are apomorphic, others not, so that it is not certain whether it is monophyletic.

According to the structure of the male genitalia, the members of the said second group fall into several subgroups, mostly not sharply differentiated, but sufficiently to be recognizable. Some subgroups consist of but a single genus (subgroups VII, VIII, IX, X, XI) with a more restricted geographical distribution. Others, such as subgroups III and VI are on the contrary composed of several genera, subgroup III being very widely distributed, being met with from North China and Japan to Australia; subgroup VI is somewhat more restricted, occurring from Indonesia to Tasmania.

The following subgroups are recognized.

### Group A.

Subgroup I. Uncus short, more or less elliptical. Tubus analis short. Valva very broad basally, then narrower, truncate terminally; harpe robust but short, placed at a right angle with the axis of the valva. Aedeagus short, rather slender. Sternal plate of the VIIIth urite large, U- or V-shaped. *Episteme* HÜBNER (China to the Philippine Islands, Java, Borneo).

### Group B.

Subgroup II. Uncus and tubus analis short. Valva variable, broad basally, or with costa and sacculus nearly parallel; harpe very short. Aedeagus with a patch of cornuti terminally; fultura inferior broad, well developed. *Asteropetes* HAMPSON, *Chelonomorpha* MOTSCHULSKY, *Maikona* MATSUMURA (Japan to North India).

Subgroup III. Uncus very long and narrow, frequently longer than the valva. Tubus analis long and narrow. Valva narrow proximally, then broad; harpe short. Aedeagus mostly short, sometimes with a long, very slender terminal process.

(a) Aedeagus without process. *Exsula* JORDAN (India-Burma), *Philippodamias* CLENCH (the Philippine Islands), *Scrobigera* JORDAN (China to the Philippine Islands and Indonesia), *Fleta* JORDAN (Indonesia), *Longicella* JORDAN (Malaya, Indonesia), *Mimeusemia* BUTLER (China and Japan to Australia), *Immetalia* JORDAN (Indonesia to New Guinea and the Solomons).

(b) Aedeagus with process. *Ophthalmis* HÜBNER (Central China to New Guinea, the Solomons, the Bismarck Archipelago, the Philippine Islands), *Crinala* JORDAN (the Philippine Islands), *Damias* WALKER (New Guinea, the Solomons, Australia), *Cruria* JORDAN (New Guinea, Australia).

Subgroup IV. Uncus moderate or short, narrow. Tubus analis moderate or long. Valva elongate; harpe short to very long. Aedeagus rather slender, sometimes with a long terminal process.

(a) Harpe short; aedeagus without process. *Cruriopsis* JORDAN (China to India), *Comocrus* JORDAN (Australia).

(b) Harpe very long; aedeagus with a long filiform process. *Crinocula* JORDAN (Borneo).

Subgroup V. Uncus moderate, hooked. Tubus analis short. Valva elongate, banana-shaped, with a terminal fringe of spinules; harpe moderate. Aedeagus slender, variously armed; fultura inferior rather narrow. *Saigonita* KIRIAKOFF (Indo-China), *Ipanica* HAMPSON (New Guinea, Australia), *Hecatesia* BOISDUVAL (Australia).

Subgroup VI. Uncus variable, short to moderately long, but always slender. Tubus analis elongate. Valva narrow to moderately broad, with a terminal fringe of spinules; harpe variable. Aedeagus slender to nearly filiform, mostly with cornuti, sometimes much broadened basally. *Argyrolepidia* HAMPSON (Indonesia to New Guinea and Australia), *Phalaenoides* LEWIN (New Guinea to Australia), *Agarista* LEACH (New Guinea to Australia), *Apina* WALKER, *Eutrichopidia* HAMPSON and *Periscepta* TURNER (Australia), *Agaristodes* HAMPSON (Tasmania).

Subgroup VII. Uncus moderately slender. Tubus analis very broad. Valva elongate, narrowed proximally and terminally, without fringe of spinules; harpe slender, rather long. Aedeagus moderate, with minute terminal cornuti; fultura inferior shaped as an inverted *V*. *Pimprana* MOORE (India).

Subgroup VIII. Uncus moderate, slender distally. Tubus analis rather narrow. Valva scimitar-shaped, with a long fringe of spinules; harpe very small, conical. Aedeagus slender, with a subterminal patch of cornuti. *Syfania* OBERTHÜR (China).

Subgroup IX. Uncus short, moderately slender, hooked. Tubus analis short and narrow. Valva narrow, with the costa and the sacculus parallel; no fringe of spinules; harpe very short. Aedeagus large and robust, with a small terminal patch of cornuti. *Aegocera* LATREILLE (India).

Subgroup X. Uncus very small, lance-shaped. Tubus analis very short. Valva moderately long, rather broad, with the costa and the sacculus more or less parallel, and without spinules; harpe very small, triangular. Aedeagus as long as the valva, rather slender, arched, with a terminal series of cornuti; fultura inferior very elongate. *Platagarista* MEYRICK (Australia).

Subgroup XI. Uncus small, slender. Tubus analis weak. Valva banana-shaped, without fringe of spinules; harpe very proximal, short, sickle-shaped. Aedeagus rather long, moderately robust, unarmed. *Radinocera* HAMPSON (Australia).

S u b g r o u p XII. Uncus rather short to moderate, slender. Tubus analis short and broad. Valva banana-shaped, sometimes broadened distally; termen with a fringe of spinules; harpe somewhat variable, short to moderate, arched or nearly straight. Aedeagus very large, moderately robust to robust, variously armed with cornuti. *Coenotoca* TURNER, *Periopta* TURNER, *Idalima* TURNER (all Australia).

S u b g r o u p XIII. Uncus moderate, rather slender. Tubus analis short. Valva banana-shaped, with very few (4–5) terminal spinules; harpe short to moderate. Aedeagus moderately robust to robust, with a terminal patch of cornuti *Sarbanissa* WALKER (China and Korea to Java and Borneo), *Zalissa* WALKER (New Guinea, Australia).

It is of course impossible to consider all the above subgroups as phyletic entities, there being nothing known about their origin and their evolution, not to speak of the possible so-called parallelisms between the genera implied. However, considering the parameters other than the male genitalia, there certainly are many affinities between the members of the subgroups. On the other hand, male genitalia of the Lepidoptera are highly complicated and elaborate structures, probably with a complicated genetic background, so that possible 'parallelisms' and 'convergences' should not be given too much prominence, especially when the space-parameter makes more probable some nearer phyletic relations between the various genera.

The author believes that the classification proposed here, even though imperfect reflects the possible evolution of the faunas somewhat better than the rather primitive existant ones.

## Artificial key to the genera

1 Antennae more or less dilated in the distal section . . . . . 2

1′ Antennae filiform, ciliate or bipectinate, not dilated distally . . . . . 26

2 Fore wings without an areole . . . . . 3

2′ Fore wings with an areole . . . . . 4

3 Palpi with the last joint long, naked; fore wings with the median vein normal . . . . . ***Episteme***, p. 8

3′ Palpi with the last joint short, hidden; fore wings with the median vein more or less dilated, ribbed . . . . . ***Saigonita***, p. 98

4 Fore wings with veins 9, 10 anastomosing with vein 8 to form an areole . . . . . 5

4′ Fore wings with vein 9 arising from vein 10, anastomosing with vein 8 to form the areole . . . . . 15

5 Hind tibiae fringed with rough hair above . . . . . 6

5′ Hind tibiae smoothly scaled . . . . . 7

6 Palpi with the last joint long, naked; abdomen clothed dorsally with rough hair . . . . . ***Comocrus***, p. 94

6′ Palpi with the last joint minute, nearly naked; abdomen smoothly scaled dorsally . . . ***Crinala***, p. 77

7 Frons with a truncate prominence with raised rim at the extremity . . . . . 8

7′ Frons with a rounded prominence . . . . . 9

8 Eyes hairy . . . . . ***Chelonomorpha***, p. 24

8′ Eyes smooth . . . . . ***Cruria***, p. 86

9 Eyes hairy . . . . . 10

9′ Eyes smooth . . . . . 11

10 Fore wings with vein 2 from towards the cell-angle . . . . . ***Crinocula***, p. 96

10′ Fore wings with vein 2 from long before the cell-angle . . . . . ***Exsula***, p. 29

11 Fore wings with vein 3 from close to the cell-angle . . . . . 12

11′ Fore wings with vein 3 from well before the cell-angle . . . . . . . . . . *Scrobigera*, p. 34
12 Fore wings with vein 2 from towards the cell-angle . . . . . . . . . . . . . . . . . . . 13
12′ Fore wings with vein 2 from long before the cell-angle . . . . . . . . . . . . . *Damias*, p. 79
13 Palpi fringed with long hair in front . . . . . . . . . . . . . . . . . . . . . . . . . 14
13′ Palpi smoothly scaled . . . . . . . . . . . . . . . . . . . . . . . . *Immetalia*, p. 60
14 Fore wings with veins 8+9, 10 from a point, at the tip of the areole . . . . . . *Fleta*, p. 42
14′ Fore wings with veins 8+9, 10 on a long stalk . . . . . . . . . . . . *Philippodamias*, p. 33
15 Eyes hairy . . . . . . . . . . . . . . . . . . . . . . . . . *Eutrichopidia*, p. 127
15′ Eyes smooth . . . . . . . . . . . . . . . . . . . . . . . . . . . . . . . 16
16 Hind tibiae spined . . . . . . . . . . . . . . . . . . . . . . . . *Syfania*, p. 134
16′ Hind tibiae not spined . . . . . . . . . . . . . . . . . . . . . . . . . . . . 17
17 Abdomen with dorsal crests on the basal urites . . . . . . . . . . . . . . . . . . . 18
17′ Abdomen without basal crests except sometimes on the first urite . . . . . . . . . . . . 21
18 Frons with a somewhat triangualr pointed prominence . . . . . . . . . . *Asteropetes*, p. 22
18′ Frons with truncate conical prominence with a raised rim at the extremity . . . . . . . . 19
19 Palpi with the second joint clothed with long hair at the extremity . . . . *Aegocera*, p. 137
19′ Palpi with the second joint evenly fringed with hair in front . . . . . . . . . . . . . . . 20
20 Abdomen with long lateral tufts of hair on the terminal urites . . . . . . . *Hecatesia*, p. 101
20′ Abdomen without lateral tufts on the terminal urites . . . . . . . . . . *Mimeusemia*, p. 47
21 Frons with a truncate rounded prominence with a raised rim at the extremity . . . . . . 22
21′ Frons with a rounded or somewhat pointed prominence . . . . . . . . . . . . . . . . . 24
22 Hind tibiae fringed with rough hair above . . . . . . . . . . . . . . *Phalaenoides*, p. 119
22′ Hind tibiae smoothly scaled . . . . . . . . . . . . . . . . . . . . . . . . . . . 23
23 Hind tibiae dilated; fore and hind wings with streaks of ribbed membrane . . . . . . . . *Platagarista*, p. 140
23′ Hind tibiae normal; wings without streaks of ribbed membrane . . . . . . . *Agarista*, p. 122
24 Pectus, femora and abdomen clothed with long hair . . . . . . . . . . *Argyrolepidia*, p. 104
24′ Pectus, femora and abdomen nearly smoothly scaled . . . . . . . . . . . . . . . . . 25
25 Both wings with vein 3 from close of the cell-angle . . . . . . . . . . . . *Ophthalmis*, p. 68
25′ Both wings with vein 3 from well before cell-angle . . . . . . . . . . . . . *Longicella*, p. 44
26 Frons with a rounded prominence with a shovel-shaped process at middle, and a corneous plate below it . . . . . . . . . . . . . . . . . . . . . . . . . . . . . . . 27
26′ Frons with the prominence not shovel-shaped and mostly without corneous plate below it . . . . . . . . . . . . . . . . . . . . . . . . . . . . . . . . . . 28
27 Eyes small, reniform . . . . . . . . . . . . . . . . . . . . . . . . . *Apina*, p. 125
27′ Eyes large, round . . . . . . . . . . . . . . . . . . . . . . . . . . *Ipanica*, p. 99
28 Frontal prominence truncate conical with raised edges . . . . . . . . . . . . . . . . 29
28′ Frontal prominence pointed or rounded . . . . . . . . . . . . . . . . . . . . . . . 36
29 Abdomen without dorsal crests . . . . . . . . . . . . . . . . . . . . . . . . . . 30
29′ Abdomen with one or more dorsal crests . . . . . . . . . . . . . . . . . . . . . . 32
30 Antennae with ciliae and setae; palpi very hairy, porrect . . . . . . . . . . *Pimprana*, p. 132
30′ Antennae simple or with ciliae only . . . . . . . . . . . . . . . . . . . . . . . . 31
31 Antennae simple; palpi shorter than the eye is broad; fore and hind wings with vein 2 from near the cell-angle . . . . . . . . . . . . . . . . . . . . . . *Periscepta*, p. 130
31′ Antennae ciliate; palpi rather long; fore and hind wings with vein 2 from well before angle . . . . . . . . . . . . . . . . . . . . . . . . . . . . . . *Idalima*, p. 150
32 Abdomen with several dorsal crests . . . . . . . . . . . . . . . . . . . . . . . . 33
32′ Abdomen with a basal crest only . . . . . . . . . . . . . . . . . . . . . . . . . 34
33 Antennae simple . . . . . . . . . . . . . . . . . . . . . . . . . . *Periopta*, p. 147
33′ Antennae ciliate . . . . . . . . . . . . . . . . . . . . . . . . . *Sarbanissa*, p. 154
34 Antennae filiform; last palpal joint dilated at the extremity . . . . . . . . . *Sarbissa*, p. 152
34′ Antennae bipectinate, serrate or ciliate; last palpal joint not dilated at the extremity . . 35
35 Antennae bipectinate in male, serrate in female; fore wings with veins 3, 4, 5 well separated; vein 6 from below the cell-angle . . . . . . . . . . . . . . *Agaristodes*, p. 128
35′ Antennae ciliate; fore wings with veins 3, 4, 5 approximated; vein 6 from the lower cell-angle . . . . . . . . . . . . . . . . . . . . . . . . . . . . *Radinocera*, p. 142
36 Frons with a pointed prominence . . . . . . . . . . . . . . . . . . . . . . . . . 37
36′ Frons with a rounded prominence . . . . . . . . . . . . . . . . . . . . . *Zalissa*, p. 168
37 Abdomen with several dorsal crests . . . . . . . . . . . . . . . . . . . *Coenotoca*, p. 144
37′ Abdomen with a basal crest only . . . . . . . . . . . . . . . . . . . . . *Maikona*, p. 27

## Genus **Episteme** HÜBNER

(fig. 1)

1820 *Episteme* HÜBNER, Verz. (12): 180. – Species typica: *Noctua lectrix* LINNAEUS (by monotypy).
1825 *Eusemia* DALMANN (syn. fide KIRBY 1892), Mon. Castniae: 26. – Species typica: *Noctua lectrix* LINNAEUS.

Diagnosis: Proboscis fully developed; eyes naked; palpi upturned, the 2nd joint fringed with long hair in front, the 3rd joint long, naked, porrect; frons with truncate prominence, with raised rim in front; antennae distally dilated towards apex; legs with smooth scaling; hind tibiae with two pairs of spurs. Wing venation: in the fore wing; vein 2 from about 5/7 of cell; 3, 4, 5 about equidistant, approximated; discocellulars strongly incurved; 7, 8, 9, 10 stalked; areole absent; 11 from cell. In the hind wing, vein 2 from 3/4 of cell; 3, 4 from angle of cell; 5 obsolescent, from middle of cell; 6, 7 short stalked or from angle of cell; 8 anastomosing with

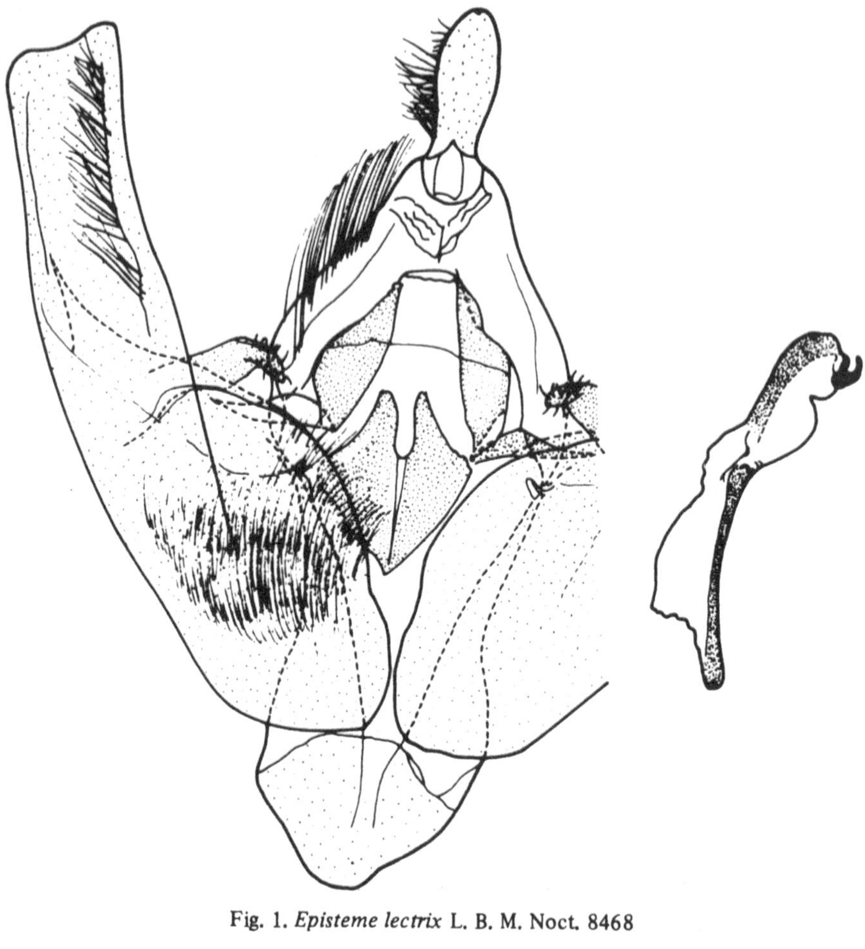

Fig. 1. *Episteme lectrix* L. B. M. Noct. 8468

the cell at a very short distance from the base. Male genitalia: Uncus rather short, lance-shaped; vinculum narrow, with the sides fused distally; the usual bunches of stiff hair absent. Valva elongate, very broad proximally, narrowing beyond the middle; apex blunt; harpe placed subcostally, short, curved, sometimes broadened terminally; fultura inferior oval, with a deep notch at middle of the distal margin. Saccus broad and short, roughly triangular. Sternal plate of the eighth urite large, U-shaped, fringed with hair.

Distribution: China, India, Indochina, Indonesia.

## Key to the species

1 Hind wings orange or red . . . . . . 2

1′ Hind wings black, usually with red or orange spots near the anal angle . . . . . . 15

2 Tegulae without yellow spots . . . . . . 3

2′ Tegulae with yellow spots . . . . . . 11

3 Collar and tegulae dotted with yellow . . . . . . *E. connexa*, p. 10

3′ Collar and tegulae dotted with white . . . . . . 4

4 Terminal band of hind wings excised and narrow in apical area . . . . . . 5

4′ Terminal band of hind wings broadening in apical area . . . . . . 10

5 Hind wings without subterminal spots . . . . . . 9

5′ Hind wings with more or less developed subterminal spots . . . . . . 6

6 Subterminal series of spots in fore wings yellow . . . . . . 7

6′ Subterminal series of spots in fore wings white . . . . . . 8

7 Fore wings with two subbasal blue lines; hind wings with the black markings less developed . . . . . . *E. hebe*, p. 12

7′ Fore wings without subbasal blue lines; hind wings with the black markings well developed . . . . . . *E. vetula*, p. 10

8 Abdomen with the VIIIth ring wholly black . . . . . . *E. darocana*, p. 13

8′ Abdomen with the VIIIth ring black above, yellow laterally . . . . . . *E. mundina*, p. 12

9 In the fore wings blue basal spots small; proximal band very much reduced . . . . . . *E. conspicua*, p. 13

9′ In the fore wings blue basal spots conspicuous; proximal band well developed, of two spots . . . . . . *E. sumatrana*, p. 13

10 Fore wings with elongate white postmedial spots; subterminal spots absent *E. bisma*, p. 14

10′ Fore wings with 1–2 yellow spots beyond the cell; subterminal series present . . . . . . *E. latimargo*, p. 14

11 Hind wings with the black terminal band not expanding between veins 2 and 4 . . . . . . 12

11′ Hind wings with the black terminal band expanding between veins 2 and 4 . . . . . . *E. lectrix*, p. 16

12 Fore wings with a series of terminal blue spots. . . . . . . *E. distincta*, p. 17

12′ Fore wings without a series of terminal blue spots . . . . . . 13

13 Anal tuft with the extremity black . . . . . . *E. nipalensis*, p. 15

13′ Anal tuft with the extremity orange . . . . . . 14

14 Fore wings with yellow spots; antemedial cell-spot white; pilosity of middle and hind tibiae black . . . . . . *E. maculatrix*, p. 19

14′ Fore wings with white or yellowish-white spots; antemedial cell-spot yellow; pilosity of middle and hind tibiae yellow . . . . . . *E. beatrix*, p. 22

15 Fore wings with the subterminal series of spots curved . . . . . . 16

15′ Fore wings with a subterminal series of obliquely placed white spots . . . *E. negrita*, p. 19

16 Abdomen with dorsal orange bands; hind wings with 2 orange spots above anal angle . . *E. adulatrix*, p. 20

16′ Abdomen without orange dorsal bands . . . . . . 17

17 Both wings with the white spots dentate towards the costa; hind wings with 2 orange spots above anal angle . . . . . . *E. sumbana*, p. 21

17′ Both wings with the white spots punctiform towards the costa; hind wings with an orange point below vein 2 . . . . . . *E. nigripennis*, p. 20

## Episteme connexa (WALKER)

1856 *Eusemia connexa* WALKER, List Lepidopt. Het. Brit. Mus., 7:1773. – Terra typica: East Java (Type, ♂, ♀, in British Museum).
1892 *Episteme connexa* – KIRBY, Syn. Cat. Lepidopt. Het.: 27.

Diagnosis: Head and thorax black; first joint of the antennae pale yellow; 1st and 2nd joints of palpi and sides of frons spotted with pale yellow; collar and tegulae with pale yellow patches; pectus with pale yellow pilosity; fore coxae yellow in front; abdomen orange, banded with bright yellow; fore wings black with yellow markings, viz. a dot below the base of costa; an antemedial cellspot; an oblique bar across middle of cell; a spot below the base of vein 2; oblique spots at both angles of cell; a curved subterminal series of small spots; and a large spot above tornus. Hind wings bright orange; base black; costa black, expanding into a spot at the upper cell-angle; a large spot beyond lower cell-angle, connected with the anal margin by an irregular band; a marginal band wide at costa, less so above vein 5, then sinuous and ending in a point at anal angle, usually with a small yellow spot on it below apex. Length of fore wing 22 mm in male, 25 in female.

Distribution: East Java.

## Episteme vetula (GEYER)

1832 *Heraclia vetula* GEYER, Zutr. Exot. Schmetterl., 4:17; fig. 657, 658. – Terra typica: Java (Type not designated).
1874 *Eusemia vetula* – BOISDUVAL, Rev. Mag. Zool. Paris, (3) 2:82.

Diagnosis: Head and thorax black; first joint of antennae white; tips of the 1st and the 2nd joints of palpi, sides of frons, a postocular streak, collar and tegulae spotted with white; abdomen orange with black bands. Fore wings black; markings pale orange or buff; a spot in middle of cell, conjoined to a spot below cell; an oblique irregular band from below costa just beyond the cell to vein 2, sometimes interrupted in discal fold, where it is bent outwards or reduced to small spots; a subterminal series of dentate spots between veins 2 and 8, and 2 others above dorsum; bluish-white subcostal marks near the base and before middle. Hind wings orange red; base black; costal area black, expanding into a spot at upper angle of cell; a large black spot beyond lower cell-angle, connected with the anal margin by an irregular band, and with the terminal band by a streak on vein 2; terminal band expanding on costa and less so above vein 5, sinuous and ending in a point at the anal angle, with a few spots on it, the largest above vein 6. Length of fore wing 30 – 33 mm.

Distribution: Assam and Burma to Indonesia and Palawan.

### Episteme vetula vetula (GEYER)

1832 *Heraclia vetula* GEYER, Zutr. Exot. Schmetterl., 4:17; fig. 657–638 (Type not designated). – Terra typica: Java.
1912 *Episteme vetula vetula* – JORDAN in SEITZ, Großschmetterl. Erde, 9:3; pl. 1c.

Distribution: Western Java.

## Episteme vetula irenea (BOISDUVAL)

1874 *Eusemia irenea* BOISDUVAL, Rev. Mag. Zool., Paris, (3) **2**:84. – Terra typica: Timor or Sumatra (Type in British Museum).
1892 *Episteme irenea* – KIRBY, Syn. Cat. Lepidopt. Het.: 27.
1912 *Eusemia maculatrix* – STRAND (nec WESTWOOD 1841), Lepidopt. Cat., **5**:5.
1912 *Eusemia vetula irenea* – JORDAN in SEITZ, Großschmetterl. Erde, **9**:3; pl. 1c.

Diagnosis: Fore wings with the subterminal series of spots white; hind wings yellower; spots on the marginal band often confluent.

Distribution: Sumatra, Nias.

## Episteme vetula fasciatrix (WESTWOOD)

1848 *Eusemia fasciatrix* WESTWOOD, Cab. Or. Entomol.: 67. – Terra typica: Assam (Type in British Museum).
1892 *Episteme fasciatrix* – KIRBY, Syn. Cat. Lepidopt. Het.: 27.
1901 *Eusemia vetula* subsp. *fasciatrix* – HAMPSON, Cat. Lepidopt. Phal. Brit. Mus, **3**:522.
1875 *Eusemia communis* BUTLER, Ann. Mag. nat. Hist., London, (4) **15**:140; tab. 13, fig. 1. – Terra typica: Assam (Type, ♂, ♀, in British Museum).
1892 *Episteme communis* – KIRBY, Syn. Cat. Lepidopt. Het.: 27.
1901 *Eusemia vetula* subsp. *communis* HAMPSON, Cat. Lepidopt. Phal. Brit. Mus., **3**:522. –
1912 *Eusemia vetula fasciatrix* forma *communis* JORDAN in SEITZ, Großschmetterl. Erde, **9**:3; pl. 1b. –

Diagnosis: Fore wings with both series of spots mostly narrower than in *E. irenea*, last spot of the antemedial series shorter and placed more distally; sometimes the spots are white (= *communis*).

Distribution: Assam, Burma; in Malaya (Penang) transition zone to *E. v. irenea*.

## Episteme vetula communicans (WALKER)

1864 *Eusemia communicans* WALKER, List Lepidopt. Het. Brit. Mus., **31**:50. – Terra typica: Cambodia (Type, ♀, in British Museum).
1892 *Episteme communicans* – KIRBY, Syn. Cat. Lepidopt. Het.: 27.
1901 *Eusemia vetula* subsp. *fasciatrix* HAMPSON, Cat. Lepidopt. Phal. Brit. Mus., **3**:522. – Terra typica: Assam (Type in British Museum).
1912 *Eusemia vetula communicans* – JORDAN in SEITZ, Großschmetterl. Erde, **9**:3.

Diagnosis: A somewhat doubtful form; transitional between *E. v. communicans* and *E. v. bijugata.*

Distribution: Cambodia.

## Episteme vetula bijugata (WALKER)

1862 *Eusemia bijugata* WALKER, J. Linn. Soc. London, Zool., **6**:85. – Terra typica: Borneo (Type in British Museum).
1892 *Episteme fasciatrix* – KIRBY, Syn. Cat. Lepidopt. Het.: 27.

1902 *Eusemia solicita* SWINHOE, Ann. Mag. Nat. Hist., London, (7) **10**:495–496. – Terra typica: Borneo (Type, ♂, in British Museum).
1912 *Eusemia vetula bijugata* – JORDAN in SEITZ, Großschmetterl. Erde, **9**:3; pl. 1b.

Diagnosis: Fore wings with the band continuous; distal band generally distinctly broader than the proximal; submarginal spots reduced or absent.

Distribution: Borneo.

## Episteme vetula palavanica (JORDAN)

1912 *Eusemia vetula palavanica* JORDAN in SEITZ, Großschmetterl. Erde, **9**:3. – Terra typica: Philippine Islands, Palawan (Type, ♂, in British Museum).

Diagnosis: Fore wings with the bands continuous, but the distal band narrower than the proximal one; submarginal series well developed; tornus with a yellow spot.

Early stages (1905 PIEPERS, Tijdschr. Ent. 's Gravenhage, **48**:195; pl. 7, fig. 6. Larva: greyish of varying intensity, darker below, with longitudinal black stripes; head, prothoracic ring, a large dorsal spot on the XIth ring and a spot above the anal legs orange.

Distribution: Philippine Isles: Palawan.

## Episteme hebe (JORDAN)

1912 *Eusemia hebe* JORDAN in SEITZ, Großschmetterl. Erde, **9**:3; pl. 4a. – Terra typica: Java (Type, ♂, in British Museum).

Diagnosis: Differing from *E. vetula* as follows. Coxae with pilosity yellow; mesothorax above with 2 lateral yellowish grey spots; metathorax above with yellowish grey lateral pilosity. Fore wings with 2 bluish lines between base and the proximal spots; submarginal spots better developed, especially the spot above tornus. Hind wings with the black markings less developed, with the transversal band shorter and the terminal band narrower.

Distribution: East and West Java.

## Episteme mundina (JORDAN)

1912 *Eusemia mundina* JORDAN in SEITZ, Großschmetterl. Erde, **9**:3, pl. 4a. – Terra typica: Perak, Mt. Tahan (Type, ♂, in British Museum).

Diagnosis: Near *E. vetula irenea*, but larger. White spots on head and thorax smaller; coxae with yellow pubescence; abdomen yellow below, unspotted or with black dots on the edge of some segments; upperside of the abdomen with the first ring broadly margined with yellow, the second with an interrupted yellow edging. Fore wings with a subbasal yellow cell-spot and two oblique broken bands, the distal one less oblique towards the dorsum than in *E. vetula*; subterminal spots white. Length of fore wing 34 – 35 mm.

Distribution: Malaya (Perak, Mt. Tahan).

### Episteme darocana (DRUCE)

1894 *Eusemia darocana* DRUCE, Ann. Mag. Nat. Hist., London, (6) **14**:23. – Terra typica: Philippine Islands (Types, ♂, ♀, in British Museum).
1901 *Eusemia vetula* subsp. *fasciatrix* HAMPSON, Cat. Lepidopt. Phal. Brit. Mus.: 522. –

Diagnosis: Head, antennae, collar and thorax black; head and collar spotted with white; abdomen dark chrome yellow, broadly banded with black; last segment black. Fore wings black; a yellow subbasal spot; two dark yellow bands across the wing, not reaching either margin; proximal band, about the middle, widest on dorsum; distal band, beyond the cell, narrow and extending almost to tornus; below end of band, close to tornus, a small pale yellow spot; a submarginal series of white dots; 4 bluish basal spots; cilia black. Hind wing brick-red, slightly paler at base and along anal margin; termen broadly margined with black; a wide black band from anal margin above tornus to end of cell; veins between the black band and the termen black. Length of fore wing 37 – 38 mm.

Distribution: Philippine Islands (Mindanao).

### Episteme conspicua ROTHSCHILD

1896 *Episteme conspicua* ROTHSCHILD, Novit. Zool. **3**:29; pl. 15, fig. 26. – Terra typica: Borneo (Types, ♂, ♀, in British Museum).
1901 *Eusemia conspicua* – HAMPSON, Cat. Lepidopt. Phal. Brit. Mus., **3**:522.

Diagnosis: Head and thorax black; antennae black with the basal joint yellow; 1st and 2nd joints of palpi, sides of frons, a postocular line, occiput, collar, tegulae spotted with yellow; pectus with yellow pilosity; coxae and pilosity of femora yellowish; abdomen orange, with black bands, narrower on underside. Fore wings black; a narrow, indistinct yellow band across middle of cell; a broad, oblique, irregular postmedial yellow band from below costa to below vein 2; a spot above vein 1; a yellow streak below it near tornus; submarginal series of spots white, if present; some blue scales near base below costa and above vein 1. Hind wings red; base black; costa black except at base; a spot at the upper cell-angle; a broad black band from just beyond lower cell-angle to anal margin, connected with the terminal band by streaks on veins 1 and 2; terminal band black, irregularly dentate on the proximal side and narrowing to the anal angle. Length of fore wing 40 – 45 mm.

Distribution: Borneo (Sarawak and North Borneo).

### Episteme sumatrana ROTHSCHILD

1899 *Episteme sumatrana* ROTHSCHILD, Novit. Zool. **6**:70. – Terra typica: West Sumatra, Padang, Sidempolan (Type, ♀, in British Museum).
1901 *Eusemia conspicua* subsp. *sumatrana* – HAMPSON, Cat. Lepidopt. Phal. Brit. Mus., **3**:522. –

D i a g n o s i s: Near *E. conspicua*. Abdomen of male spotted with black on the underside; mesothorax spotted with bluish. Fore wings: a yellow dot at base of cell; transverse bands broad, orange yellow, frequently fused together dorsally; a subterminal series of white dots, comprising the dot above tornus; blue basal marks larger.

D i s t r i b u t i o n: West Sumatra: Padang Bovenlanden.

## Episteme bisma (MOORE)

1859 *Eusemia bisma* MOORE, Lepidopt. East Ind. Co.: 287. – T e r r a t y p i c a: Java (Types, ♂, ♀, in British Museum).
1874 *Eusemia lambertiana* BOISDUVAL, Rev. Mag. Zool., (3) **2**:83. – T e r r a t y p i c a: "Timor ou Sumatra" (Type in British Museum).
1892 *Episteme bisma* – KIRBY, Syn. Cat. Lepidopt. Het.: 27.

D i a g n o s i s: Antennae, head and thorax black; 1st and 2nd joints of palpi, sides of frons, postocular line, collar, tegulae and prothorax with white dots; pectus with yellow pilosity; fore coxae yellow in front; abdomen orange yellow, with broad black dorsal bands and narrow ventral lines. Fore wings black; an orange point in proximal part of cell; a square spot in middle of cell, with a small dot above it, and a triangular spot below base of vein 2; a curved postmedial row of elongate white spots between veins 1 and 8; a bluish subbasal band. Hind wings red; base black; costa black, expanding into a spot on upper cell-angle; 3 large conjoined spots between lower cell-angle and anal margin, connected with the terminal band by streaks on veins 1 and 2; black terminal band with irregular proximal margin, expanding widely apically, narrowing to a point at anal angle, and usually with white spots on it above veins 2, 3 and 6. Length of fore wing 36 – 40 mm.

E a r l y s t a g e s (MOORE, 1859 Lepidopt. East Ind. Co. 287). Larva black, with yellow or white intersegmental bands; each ring with a transversal row of shiny black tubercles bearing each a long whitish hair; head black or orange spotted with black; legs black. On *Dioscorea oppositifolia* and on *D. batatas*.

D i s t r i b u t i o n: Java, Sumatra, Bali, Lombok, Soembawa, Soemba.

## Episteme latimargo (HAMPSON)

1891 *Eusemia latimargo* HAMPSON, Illustr. Het. Brit. Mus., **8**:46; pl. 24. – T e r r a t y p i c a: Western slopes of the Nilghiris (Types, ♂, ♀, in British Museum).
1892 *Episteme latimargo* – KIRBY, Syn. Cat. Lepidopt. Het.: 27.
1912 *Eusemia latimargo* var. *shevaroyensis* STRAND, Lepidopt. Cat., **5**:4. – T e r r a t y - p i c a: Shevaroy Hills (Type, ♀, in British Museum).

D i a g n o s i s: Head and thorax black; 1st joint of antennae white; 1st and 2nd joints of palpi, sides of frons, postocular line, collar, tegulae and mesothorax marked with white; fore coxae orange in front; mid and hind coxae with orange pilosity; abdomen bright orange with dorsal black bands and lateral and ventral series of spots. Fore wings black; a square pale yellow spot at middle of cell, with a streak above it; a similar spot below base of vein 2; discocellulars with several dots; a curved subterminal series of dentate white spots; 2 blue spots below base of costa and 2 above vein 1. Hind wings carmine red or orange; base black; costal area black, emitting a triangular spot at upper cell-angle; 2 large confluent spots between lower cell-angle and anal margin, connected with the terminal band by streaks on veins 1 and 2; black terminal band very wide on apical area to vein 4, narrowing to a point at anal angle; a white spot on it above vein 6 and usually another above vein 8. Length of fore wing 37 – 42 mm.

Distribution: South India (Malabar, Nilghiris, Travancore).

## Episteme nipalensis (BUTLER)

1848 *Eusemia maculatrix* WESTWOOD, Cab. Or. Entomol.: 67; pl. 33, fig. 1 (nec WESTWOOD 1841, fide HAMPSON 1901).
1875 *Eusemia nipalensis* BUTLER, Ann. Mag. Nat. Hist., London, (4) **15**:140. – Terra typica: Nepal (Type in British Museum).
1881 *Eusemia dives* BUTLER, Illustr. Het. Brit. Mus., **5**:18; pl. 82, fig. 4. – Terra typica: Sikkim (Type, ♂, ♀, in British Museum).
1892 *Episteme nipalensis* – KIRBY, Syn. Cat. Lepidopt. Het.: 27.
1897 *Episteme westwoodi* KIRBY, Handb. Lepidopt., in ALLEN's Nat. Libr., **3**:65. – Terra typica: non designata. – Terra typica restricta: Nepal.
1912 *Eusemia nipalensis* aberr. *clara* JORDAN in SEITZ, Großschmetterl. Erde, **9**:4. – Terra typica: Assam, Khasia Hills (Type in British Museum).
1939 *Eusemia nipalensis* aberr. *semiclara* BYTINSKI–SALZ, Entomolog. Rec. J. Var. London, **1939**:117.
1939 *Eusemia nipalensis* f. *hainana* BYTINSKI–SALZ, Entomolog. Rec. J. Var. London, **1939**:117. – Terra typica: Hainan.

Diagnosis: Head and thorax black; 1st joint of antennae white; 1st and 2nd joints of palpi, sides of frons, postocular line marked with white; collar and tegulae marked with yellow; coxae abd femora with orange pilosity; abdomen orange with black dorsal bands, lateral and ventral spots, and the extremity. Fore wings black; markings yellow, sometimes white; a large spot at middle of cell, and another below base of vein 2; 3 small confluent spots beyond each cell-angle; a curved subterminal series of dentate white spots, spot in interval II sometimes absent. Hind wings carmine red or orange; base black; costal area black, emitting a triangular spot at upper cell-angle; a band of two confluent spots from lower cell-angle to anal margin, connected with the terminal band by streaks on veins 1 and 2; terminal band very wide apically to vein 4, narrowing to a point at the anal angle; a white subapical spot in it, sometimes accompanied by others; also spots in intervals II and III. Length of fore wing 33 – 44 mm.

Biology: In North India flies at altitudes from 1500 to 2000 m (in April–May and in August).

Distribution: North India to Burma and Vietnam.

### Episteme nipalensis nipalensis (BUTLER)

1912 *Eusemia nipalensis nipalensis* — JORDAN in SEITZ, Großschmetterl. Erde, **9**:4; pl. 1a, 1b.

Distribution: Nepal (Terra typica), Sikkim, Assam.

## Episteme nipalensis expansa (JORDAN)

1912 *Eusemia nipalensis expansa* JORDAN in SEITZ, Großschmetterl. Erde, **9**:4. – Terra typica: Tonkin, Chien-Hoa (Type, ♂, in British Museum).

Diagnosis: The four large spots of fore wings yellow; spot below cell larger than the spot in cell and than the lower antemedial spot.

Distribution: Vietnam (in April to July and in August to September).

## Episteme nipalensis schana (JORDAN)

1912 *Eusemia nipalensis schana* JORDAN in SEITZ, Großschmetterl. Erde, **9**:4. – Terra typica: Tenasserim (Type, ♂, in British Museum).

Diagnosis: Fore wings somewhat narrower. Discal band on hind wings narrower than space between it and the marginal band.

Distribution: Tenasserim.

## **Episteme lectrix** (LINNAEUS)

1764 *Noctua lectrix* LINNAEUS, Mus. Ulr.: 389. – Terra typica: "China" (Type not designated). – Terra typica restricta: South China, Kwangtung (MELL 1936).
1781 *Bombyx lectrix* – FABRICIUS, Spec. Ins., **2**:201.
1854 *Eusemia lectrix* – WALKER, List. Lepidopt. Het. Brit. Mus., **1**:47.
1892 *Episteme lectrix* – KIRBY, Syn. Cat. Lepidopt. Het.: 26.
1939 *Eusemia lectrix* aberr. *reducta* BYTINSKI–SALZ, Entomolog. Rec. J. Var., **1939**:119. – Terra typica: China.

Diagnosis: Head and thorax black; basal joint of antennae white; 1st and 2nd joints of palpi, sides of frons, postocular line white; collar spotted pale yellow; tegulae spotted with yellow; fore coxae orange yellow in front; abdomen orange with black bands. Fore wings black; a small yellow antemedial spot in cell; a square spot in middle of cell, with two short streaks above it, and a square spot below base of vein 2; 5 dots beyond angles of cell; a curved subterminal series of small white spots; spot in interval II often absent; 4 blue basal spots. Hind wings black, with an orange antemedial band not reaching costa, and enclosing an elliptical black spot at upper cell-angle, where it is confluent with a double dentate spot beyond cell; 2 spots beyond lower cell-angle, and 2 above and below vein 1 towards the anal angle; subterminal white spots in intervals II and III and above and below vein 6. Length of fore wing 28 – 45 mm.

Early stages: Larva brownish, with several longitudinal rows of black spots; each segment at middle with a raised ring bearing a number of minute black warts, each with a long greyish-white hair; head, pronotum and tergite of penultimate ring, and legs shiny black. On *Dioscorea batatas*, *Smilax china*.

Distribution: West, Central and South China including Taiwan.

### Episteme lectrix lectrix (LINNAEUS)

1936 *Episteme lectrix lectrix* – MELL, Stettin. Entomol. Ztg., **97**:30.

Distribution: South China (Kwangtung, possibly Fukien).

### Episteme lectrix szechuanensis (MELL)

1936 *Eusemia lectrix szechuanensis* MELL, Stettin. Entomol. Ztg., 97:29. – Terra typica: China, Szechuan.

Diagnosis: Abdomen ochreous brown, with broad bands, fore wings with the subbasal spot as large as the spot below cell; both antemedian spots fused; hind wings ochreous brown in median area.

Distribution: Szechuan.

### Episteme lectrix hoenei (MELL)

1936 *Eusemia lectrix hönei* MELL, Stettin. Entomol. Ztg., 97:29. – Terra typica: China, Chekiang.

Diagnosis: Abdomen deep yellow; bands broad, underside spotted in male, banded in female; fore wings with the subbasal spot smaller than the spot below cell; hind wings deep yellow.

Distribution: Chekiang.

### Episteme lectrix sauteri (MELL)

1936 *Eusemia lectrix sauteri* MELL, Stettin. Entomol. Ztg., **97**:30. – Terra typica: China (Taiwan).

Diagnosis: Abdomen orange yellow, with large bands and spots; fore wings with the markings small; upper antemedian spot much smaller than the lower one; subbasal cell-spot as large as or larger than the 2nd spot in the postmedian series.

Distribution: Taiwan.

### **Episteme distincta** (BUTLER)

1875 *Eusemia distincta* BUTLER, Ann. Mag. Nat. Hist., London, (4) **15**:140. – Terra typica: Assam, Khasis (Types, ♂, ♀, in British Museum).
1892 *Episteme distincta* – KIRBY, Syn. Cat. Lepidopt. Het.: 27.
1901 *Eusemia maculatrix* aberr. *distincta* – HAMPSON, Cat. Lepidopt. Phal. Brit. Mus., **3**:524. –

Diagnosis: Near to *E. nipalensis* (BUTLER). Pronotum unspotted; spots on tegulae very large; first abdominal segment without yellow rim above; eighth segment without yellow rim above; eighth segment yellow, black at base in male;

extremity of abdomen yellow in female; femora black; fore coxae yellow in front, mid and hind coxae yellow at extremity; tibiae frequently streaked with white above. Fore wings with the blue basal spots large; medial spots yellow, almost conjoined; postmedial spots white; second discal spot composed of a square patch, with dots before and behind it, which frequently are absent; a subterminal series of blue streaks, always absent in *E. nipalensis*, *E. maculatrix* and *E. lectrix*. Hind wings yellow, with the medial black patch not connected with the costa, but isolated and nearly circular.

Distribution: Silhet, Assam, Sikkim, North Burma, China.

### Episteme distincta distincta (BUTLER)

Distribution: India and Burma.

### Episteme distincta tatsienlouica (BYTINSKI–SALZ)

1939 *Eusemia distincta tatsienlouica* BYTINSKI–SALZ, Entomolog. Rec., **51**:118; pl. 9, fig. 3 (♂). – Terra typica: Tatsienlou.

Diagnosis: Fore wings with the median spots larger and the postmedian spots reduced; spots yellow, not white; hind wings black, without marginal streaks.

Distribution: China: Tatsienlou.

### **Episteme macrosema** (JORDAN)

1912 *Eusemia distincta* subsp. *macrosema* JORDAN in SEITZ, Großschmetterl. Erde, **11**:5. – Terra typica: Momeit (Type, ♀, in British Museum).
1920 *Eusemia macrosema* – HAMPSON, Cat. Lepidopt. Phal. Brit. Mus., Suppl. **2**:535; pl. 70, fig. 1 (♂).

Diagnosis: Head and thorax black; 1st and 2nd joints of palpi and sides of frons with white dots; tegulae spotted with pale yellow; fore and mid tibiae and first joint of fore tarsi streaked with white; abdomen banded with orange red and black; anal tuft black; underside orange with lateral and subventral series of black spots. Fore wings black; a broad, oblique, yellowish white medial band with slightly irregular edges, from below costa to vein 1; a white dot beyond the cell in interval VI, and a larger one in interval III; an oblique series of 4 white elongate spots, subterminally in intervals IV–VII, the two lower obsolete, and a small spot in submedian fold; basal blue spots very strongly reduced. Hind wing orange red; base, costal and terminal areas black; a black spot on upper discocellular conjoined to the costal area; a large, oblique, irregularly elliptical black patch from and from beyond lower cell-angle to tornus, connected to the terminal area by streaks on veins 1 and 2; terminal black area very broad at costa, bearing a small subterminal white spot in interval VI and a minute dot in interval III. In female the extremity of the abdomen is orange, and the fore wings bear 3 small spots beyond the upper cell-angle. Length of forewing 32 – 34 mm.

Distribution: Burma (Terra typica: Momeit), Central and West Yunnan.

### Episteme maculatrix (WESTWOOD)

1841 *Eusemia maculatrix* WESTWOOD, in JARDINE, Nat. Libr. 32 Exot. Moths, 5:88; pl. 2, fig. 3. – Terra typica: Assam (fide JORDAN 1913). (Type missing).
1891 *Episteme maculatrix* – KIRBY, Syn. Cat. Lepidopt. Het.: 26.

Diagnosis: Head and thorax black; first antennal joint white; 1st and 2nd joints of palpi, sides of frons, postocular streak marked with white; tegulae with large yellow spots; pronotum unspotted; legs black, with only coxal orange pilosity; first abdominal segment black with an yellow spot; eighth segment orange with a black basal spot; extremity orange in both sexes; rest of abdomen orange with black bands and lateral and ventral series of spots. Fore wings black; a small antemedial white spot in cell; a square white spot in middle of cell, with 2 streaks above it; a separated spot below base of vein 2; discal spots small, white, 3 of them beyond upper angle of cell, the middle one the largest and square, and 1–2 beyond lower angle of cell; blue spots present in basal area only. Hind wings deep yellow; costal area black; an elliptical spot at the upper angle of cell, connected to costa; 2 confluent spots between lower cell-angle and anal margin, connected by streaks on veins 1 and 2 with the terminal band; latter wide on apical area to vein 5, narrowing to a point at anal angle, with a white spot in interval VI, and sometimes with additional smaller spots between intervals VI and II. Length of fore wing 28 – 41 mm.

Distribution: Assam.

### Episteme negrita (HAMPSON)

1894 *Eusemia negrita* HAMPSON, Moths India, 2:151. – Terra typica: Burma, South Shan States, Enlay (Type, ♂, in British Museum).

Diagnosis: Head and thorax black; palpi marked with white laterally; tegulae spotted with yellow; coxae with orange pilosity; fore and mid tibiae and tarsi with some white in front; abdomen black with orange red bands on 2nd, 3rd and 7th segments; underside orange red with black transverse spots. Fore wings black; a broad oblique antemedian band of yellow from below costa to vein 1, with irregular edges; traces of a white dot beyond the upper cell-angle, and a more distinct spot beyond the lower angle; 5 postmedial points in an oblique row from below costa to interval III, with traces of a point in interval I; 3 blue basal spots and a few blue scales. Hind wings black with slight subbasal orange red streaks in cell, on media and on anal margin, a subterminal spot in interval I and small bluish white spots in intervals III and VI. Length of fore wing 29 mm.

Distribution: Burma: Shan States, Enlay.

## Episteme nigripennis (BUTLER)

1975 *Eusemia nigripennis* BUTLER, Ann. Mag. Nat. Hist., London, (4) **15**:140. – Terra typica: Ceylon (Types, ♂, ♀, in British Museum).
1882 *Episteme nigripennis* – MOORE, Lepidopt. Ceylon, **2**:34; pl. 94, fig. 3.

Diagnosis: Head and thorax black; first antennal joint white; 1st and 2nd joints of palpi, sides of frons, postocular streak spotted with white; collar and tegulae spotted with yellow; abdomen black; 1–2 terminal segments ringed with orange in female; underside banded with orange, except at base. Fore wings black; a square yellow cell-patch to below costa, confluent with a patch below cell, extending to above base of vein 1; small spots beyond cell in intervalls III–VI, and a slight streak below costa; a curved subterminal series of white points, with small blue spots below costa and above dorsum; base with bluish spots. Hind wings black, with an obliquely curved postmedial series of small bluish-white spots from costa to vein 2, and an orange spot in interval I. Length of fore wing 34 – 35 mm.

Distribution: Ceylon.

## Episteme adulatrix (KOLLAR)

1844 *Eusemia adulatrix* KOLLAR, HÜGEL's Kaschmir, **4** (2):464; pl. 20, fig. 1. – Terra typica: Kaschmir (Type not designated).
1892 *Episteme adulatrix* – KIRBY, Syn. Cat. Lepidopt. Het.: 26.
1939 *Eusemia adulatrix* aberr. *postnigra* BYTINSKI–SALZ, Entomol. Rec., **51**: pl. 9, fig. 5. –

Diagnosis: Head and thorax black; first antennal joint white; 1st and 2nd joints of palpi and sides of frons with white points; a white postocular line; collar, tegulae and prothorax spotted with yellow; inside of fore coxae orange; abdomen black, with orange red bands in distal area, at least three bands in both sexes; underside with a complete series of bands, except on basal segment. Fore wings black; a large square yellow patch in middle of cell; a similar patch below cell, extending to just above base of vein 2; an oblique half-band beyond cell, with spots in intervals III–VI, and two streaks on costa; a curved subterminal series of white dots, except in interval IV bluewhite basal spots, subbasal and antemedial lines from cell to a streak above dorsum, often reduced; a dorsal blue streak, sometimes absent. Hind wings black, with postmedial white spots in intervals II, III and VI, and orange red spots above anal angle. Length of fore wing 30 – 36 mm.

Distribution: Cashmir, Pakistan, India, to Burma, Indo-china, Hainan, N. W. Yunnan, Szechuan, Kweichow, Kwangtung.

### Episteme adulatrix adulatrix (KOLLAR)

1848 *Eusemia bellatrix* WESTWOOD, Cab. Or. Ent.: 67; pl. 33, fig. 2. – Terra typica: India (Type, ♂, in Mus. Oxon.).
1875 *Eusemia sectinotis* BUTLER, Entomologist's mon. Mag., **12**:117. – Terra typica: Punjab, Mussoorie (Types, ♂, ♀, in British Museum).
1912 *Eusemia adulatrix adulatrix* – JORDAN in SEITZ, Großschmetterl. Erde, **11**:6; pl. 1c.

D i a g n o s i s: As above; rather variable.

D i s t r i b u t i o n: Cashmir to Shan States, Vietnam, West China.

### Episteme adulatrix contracta (BUTLER)

1875 *Eusemia contracta* BUTLER, Entomologist's mon. Mag., **12**:117. – T e r r a t y p i c a: Punjab, Mussoorie (Types, ♂, ♀, in British Museum).
1875 *Eusemia afflicta* BUTLER, Entomologist's mon. Mag., **12**:117. – T e r r a t y p i c a: Bombay (Type, ♂, in British Museum).
1875 *Eusemia simplex* BUTLER, Entomologist's mon. Mag., **12**:118. – T e r r a t y p i c a: Kanara (Type, ♂, in British Museum).

D i a g n o s i s: Blue markings in basal area of fore wings reduced. Abdominal bands and anal mark on hind wings orange yellow.

D i s t r i b u t i o n: South India: Nilghiris, Kanara, Poona, Mhow, Bombay.

### Episteme adulatrix spilosa (JORDAN)

1912 *Eusemia adulatrix* subsp. *spilosa* JORDAN in SEITZ, Großschmetterl. Erde, **11**:6. – T e r r a t y p i c a: Hainan, Cheng-Mai (Type, ♂, in British Museum).

D i a g n o s i s: Medial spots on fore wings reduced and well separated; submarginal pale spots well developed, except the 4th and the 6th, which are mostly small, the 4th being often absent.

D i s t r i b u t i o n: Hainan.

### **Episteme sumbana** (ROTHSCHILD)

1897 *Episteme sumbana* ROTHSCHILD, Novit. Zool., **4**:310. – T e r r a t y p i c a: Soemba Island (Type, ♂, in British Museum).
1901 *Eusemia sumbana* – HAMPSON, Cat. Lepidopt. Phal. Brit. Mus., **3**:527; pl. 51, fig. 1.

D i a g n o s i s: Body black; 1st and 2nd joints of palpi and sides of frons with white dots; collar and tegulae with pale yellow spots; coxae black; fore coxae with yellow scales on inner side; abdomen with lateral spots and ventral bands on urites IV–VII. Fore wings black; square pale yellow patches at extremity of cell and below base of vein 2; a dot above base of vein 2; just beyond the cell, a series of five small spots; a curved subterminal series of seven small white spots, more proximal than in other species of the genus; subbasal and antemedial oblique bluish broken lines. Hind wings black with an oblique series of six small elongate white spots just beyond the cell, and two orange red spots above anal angle; base with an orange tinge by transparency from the underside. Length of fore wing 33 mm.

D i s t r i b u t i o n: Soemba Island.

## **Episteme beatrix** (JORDAN)

1899 *Eusemia irenea* – LEECH, Trans. Entomol. Soc., **1899**:211, nec BOISDUVAL 1874.
1909 *Eusemia beatrix* JORDAN in SEITZ, Großschmetterl. Erde, **3**:6; pl. 1a (♂). – Terra typica: West China, Omeishan (Type, ♂, in British Museum).
1912 *Eusemia maculatrix* aberr. *antemedialis* STRAND, Lepidopt. Cat., **5**:5. – (aberr. 1 of HAMPSON, Cat. Lepidopt. Brit. Mus., **3**:527).
1912 *Eusemia maculatrix* aberr. *cellularis* STRAND, Lepidopt. Cat., **5**:5. – (aberr. 3 of HAMPSON, Cat. Lepidopt. Brit. Mus., **3**:527).
1912 *Eusemia maculatrix* aberr. *angularis* STRAND, Lepidopt. Cat., **5**:5. – (aberr. 4 of HAMPSON, Cat Lepidopt. Brit. Mus., **3**:527).
1914 *Episteme beatrix* var. *formosana* MATSUMURA, Thous. Ins., Suppl. **2**:72. – Terra typica: Formosa, i. e. China, Taiwan.

Diagnosis: Head and thorax black; 1st antennal joint yellow; 1st and 2nd joints of palpi spotted with white; sides of frons with yellow dots; tegulae spotted with yellow; pectus and flanks orange; fore tibiae streaked with white at base; mid and hind tibiae with yellowish pilosity; abdomen orange with dorsal black bands, lateral spots and ventral spots on terminal segments, in female extremity unspotted below; last joint orange; pilosity of valvae in male black. Fore wings black; an antemedial yellow cell-spot; in and below the cell square median yellow patches; three small postmedial yellow spots below costa and three white spots between veins 2 and 5, the middle one larger; four small wedge-shaped subapical white spots and a point above tornus; subbasal and antemedial blue marks below costa and cell. Hind wings orange; some black at base; a discoidal spot; a patch from lower cell-angle to just above the anal margin, connected by streaks on veins 1 and 2 with the terminal band, expanding into a square patch between costa and vein 4, and with a white subapical spot on it. Length of fore wing 29 – 36 mm.

Distribution: West China: Szechuan, Hunan, Chekiang, also Taiwan.

## **Genus Asteropetes** HAMPSON

(fig. 2)

1901 *Asteropetes* HAMPSON, Cat. Lepidopt. Phal. Brit. Mus., **3**:605–606. – Species typica: *Seudyra noctuina* BUTLER (by original designation).

Diagnosis: Proboscis fully developed; antennae weakly dilated beyond middle; palpi upturned; 1st and 2nd joints with long pilosity; 3rd long, naked, dilated at extremity, porrect; frons with a beak-like triangular prominence; tibiae fringed with long hair; abdomen with large basal crests. Venation: in fore wings, vein 2 from about three-quarters of cell; 3 and 4 approximated; 5 well above the angle of cell; distance 4–5 about twice that of 3–4; 6 from the upper angle of cell; areole narrow; 7 and the long stalk of 8+9 from tip of areole; 10 from upper side of areole near the extremity. In hind wings, vein 2 from slightly beyond two-thirds of cell; 3, 4 and 6, 7 from cell-angles; discocellulars incurved; 5 obsolescent, from middle of discocellulars.

Male genitalia: Uncus small, narrow, rather short, about one-fourth of sacculus, with sparse pilosity and a small terminal hook. Tubus analis short, sclerified.

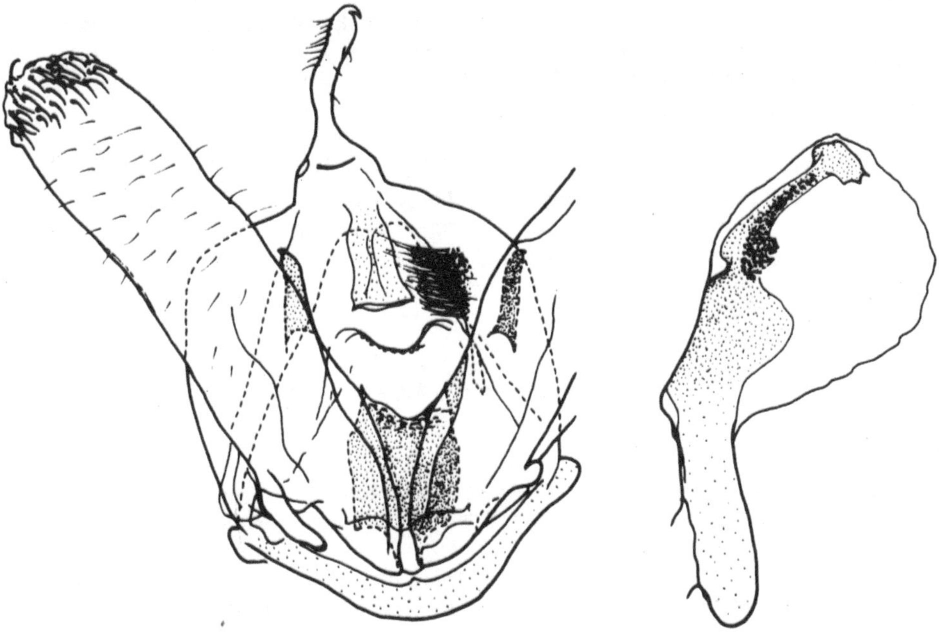

Fig. 2. *Asteropetes noctuina* (BUTLER). Type B. M. Noct. 8392

Vinculum broad, bearing before the distal extremity a fringe of stiff hair. Valva elongate, with costa and sacculus parallel; termen rounded, bearing several rows of short spinelike hairs; harpe short, slender, with a small terminal hook; sides of harpe minutely serrate; harpe placed at the proximal half of valva, nearer costa than sacculus. Aedeagus slightly longer than three-quarters of the sacculus, rather robust, spoon-shaped in the proximal portion; extremity produced into a narrow process ending in a toothed plate, and partly covered with minute cornuti; fultura inferior well developed, with the sides faintly excurved, the distal angles produced and pointed, and the terminal margin incurved. Saccus very short and broad, outcurved.

D i s t r i b u t i o n: Japan.

### Asteropetes noctuina (BUTLER)

1878 *Seudyra noctuina* BUTLER, Entomologist's mon. Mag., **14**:206. – T e r r a t y p i c a: Hakodate, Japan (Types ♂, ♀, in British Museum).
1901 *Asteropetes noctuina* – HAMPSON, Cat. Lepidopt. Phal. Brit. Mus., **3**:606; fig. 264 (♂, ♀).

D i a g n o s i s: Head and thorax mixed black and grey; antennae and palpi blackish; mid and hind tibiae fringed with orange hair; abdomen orange, with dorsal crests blackish mixed with grey and rufous at base; anal tuft black; underside grey. Fore wings reddish-violet-brown; below cell, a brown streak emitting three oblique lines to vein 1; spots in cell and on discocellulars black with some brown scales, and edged with grey; discocellular spot with a yellowish spot beyond it, emitting three rays not reaching the subterminal line; costa above browner; subterminal line black, double, filled in and edged on both sides with brown; terminal area purplish carmine, edged with red proximally and distally, the distal edge with black and white scales proximally in the tornal area; cilia brown. Hind wings orange with a round black discocellular spot and a broad black terminal band, extending to the anal margin; an orange terminal spot near tornus; cilia yellowish white with a brown median line. Length of fore wing 19 mm.

Distribution: Japan.

## Genus Chelonomorpha MOTSCHULSKY

(fig. 3)

1860 *Chelonomorpha* MOTSCHULSKY, Études entomologiques, Helsingfors, 9:30. – Species typica: *Chelonomorpha japona* MOTSCHULSKY (by monotypy).

Diagnosis: Proboscis fully developed; antennae distinctly dilated towards the extremity; palpi upturned; 1st and 2nd joints fringed with long hair in front; 3rd joint smooth, porrect; eyes with very short pilosity; tibiae and tarsi with smooth hair and scales; abdomen of male with protrusible lateral tufts of long hair from the

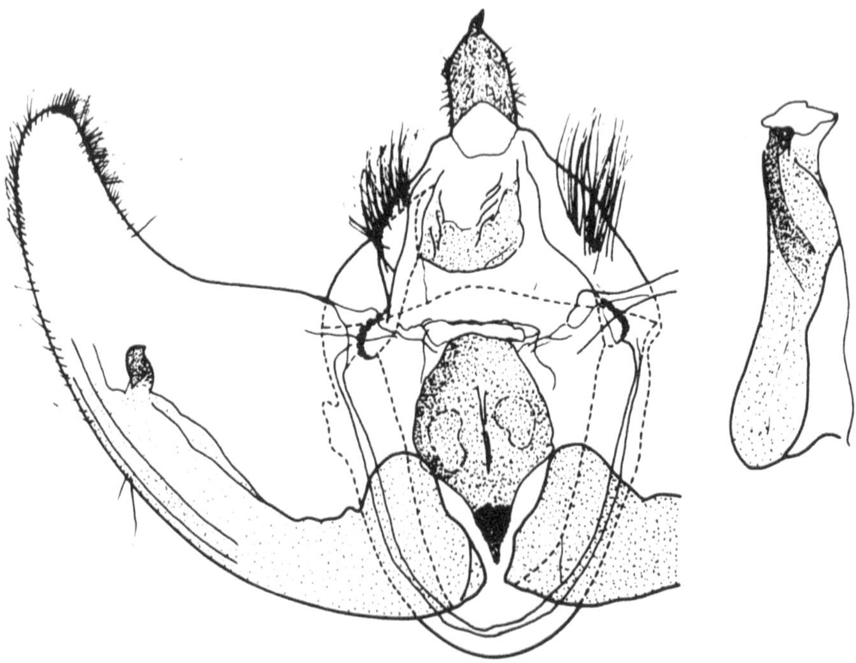

Fig. 3. *Chelonomorpha japona* MOTSCHULSKY. Type of *Eusemia villicoides* Butler. B. M. Noct. 8469

base. Nervation: Fore wing with vein 2 from 2/3 of cell; 3, 4 and 5 approximated, about equidistant; 6 from just below the upper cell-angle; discocellulars strongly incurved; an areole present; 7 and (8 + 9) and 10 from tip of areole; 11 from cell. Hind wing with vein 2 from 3/4 of cell; 3, 4 and 6, 7 from angles; 5 obsolescent, from middle of the discocellulars; 8 anastomosing at a point or for a short distance with cell, near the base.

Male genitalia: Uncus small, lance-shaped. Vincula broad, fringed with stiff hair. Valva very broad basally, becoming narrower and tapering to a subacute apex, strongly arched; basal fold of sacculus nearly circularly expanded basally, then rather narrow, reaching to just beyond the middle of the valva, and ending in a small, short harpe directed costalwards. Aedeagus about 1/2 of valva in length, club-shaped proximally, slightly broadened distally; fultura inferior broad-oval, with the proximal margin produced, pointed and covered with minute spinules. Saccus short, rounded.

D i s t r i b u t i o n: Japan, China, North India.

Key to the species

1 Coxae black . . . . . . . . . . . . . . . . . . . . . . . . . . . . . . . . . . . 2
1′ Coxae yellow . . . . . . . . . . . . . . . . . . . . . . . . . . *Ch. austeni*, p. 26
2 Fore wings with the square spots small, pure white; hind wings with the apical spot white . . . . . . . . . . . . . . . . . . . . . . . . . . *Ch. formosana*, p. 26
2′ Fore wings with the square spots larger, yellowish; hind wings with the apical spot orange . . . . . . . . . . . . . . . . . . . . . . . . . . *Ch. japona*, p. 25

## Chelonomorpha japona (MOTSCHULSKY)

1860 *Chelonomorpha japona* MOTSCHULSKY, Études entomologiques, Helsingfors, 9:29. – T e r r a t y p i c a: Jesso, Japan (Type not designated).

1875 *Eusemia villicoides* BUTLER, Ann. Mag. Nat. Hist., (4) 15:141; pl. 13, fig. 2 (♂, ♀). – T e r r a t y p i c a: Japan. (Types, ♂, ♀, in British Museum).

D i a g n o s i s: Head, body and appendages black; head and collar with white dots; tegulae yellow; coxae sometimes with whitish hair; abdomen orange with black markings, viz. dorsal and lateral bands and sometimes also ventral spots. Fore wings black with yellowish markings: a triangular cell-spot; a broad band at extremity of cell with a square spot below; an interrupted bar beyond the cell, consisting of square spots of variable size; a subterminal series of small spots or dots; a few basal silvery-blue spots; an oblique postmedian line from vein 3 to near dorsum; an oblique cell-spot and a discoidal band, all silvery-blue. Hind wings orange; markings black: a basal spot; a costal streak; a small spot on upper discocellular; a large spot below lower discocellular; a spot in the anal area; a terminal band, much wider apically, tapering towards the anal angle, mostly with one or more orange spots. Length of fore wing 28 – 33 mm.

D i s t r i b u t i o n: Japan and China to Burma.

E a r l y s t a g e s: Larva leather-brown, with black warts, each bearing a dark hair; head black. Possibly on *Trema amboinensis* and *Adinandra milletti*.

### Chelonomorpha japona japona (MOTSCHULSKY)

Distribution: Japan, Korea, North and Central China.

### Chelonomorpha japona dubia (JORDAN)

1912 *Chelonomorpha dubia* JORDAN in SEITZ, Großschmetterl. Erde, **11**:6 (♂). – Terra typica: Hsipaw, Burma (Type, ♂, in British Museum).
1912 *Chelonomorpha japona* var. *burmana* STRAND, Lepidopt. Cat., **5**:6. – Terra typica: Hsipaw, Burma (Type ♂, in British Museum).

Diagnosis: Fore coxae with whitish pilosity. Abdomen unspotted laterally and below.

Distribution: Burma.

### Chelonomorpha japona kansuana (BYTINSKI-SALZ)

1939 *Chelonomorpha japona kansuana* BYTINSKI–SALZ, Entomologist's Rec. J. Variat., London, **51**:119; pl. 9, figs. 7, 8 (♂, ♀). – Terra typica: China, SE Kansu (Types, ♂, ♀, in coll. BYTINSKI–SALZ).

Diagnosis: Fore wings with the yellowish spots somewhat reduced, the two postmedial spots separated; hind wings with the black spots beyond the cell conjoined.

Distribution: Southeastern Kansu.

### Chelonomorpha austeni (MOORE)

1879 *Eusemia austeni* MOORE, Lepidopt. Atk.: 11 (♂). – Terra typica: Khasis, Assam (Type, ♂, in Museum Berlin).
1894 *Chelonomorpha japona* var. *austeni* – HAMPSON, Moths Ind., **2**:154.

Diagnosis: Differing from *Chelonomorpha japona* as follows: Fore coxae clothed with long yellow hair; spots on sides and underside of abdomen small. Pale spots on fore wings smaller; subbasal spot rounded; post-cellular series shorter. Hind wings with the spots below lower discocellular and near the anal margin fused, sometimes connected with the terminal band. Length of fore wing 20 – 27 mm.

Distribution: Assam (Terra typica); Central and South China (Hupé, S. Shensi, Taipaishan).

### Chelonomorpha formosana (MIYAKE)

1907 *Chelonomorpha formosana* MIYAKE, Annot. Zool. Jap., Tokyo, **6**:79 (♂). – Terra typica: Taikokan, Taiwan.
1912 *Chelonomorpha japona formosana* – JORDAN in SEITZ, Großschmetterl. Erde, **11**:6.

D i a g n o s i s: Differing from *Chelonomorpha japona* as follows: Fore wings with the silvery-blue spots much broader and more conspicous; an antemedial oblique line from costa to dorsum; some streaks on vein 1 and on media; square patches much smaller and pure white instead of yellowish. Apical patch on the hind wings also white. Fore wings on underside with eight marginal spots as against four in *Ch. japona*; hind wings on underside with only four marginal spots. Length of fore wing 26 – 30 mm.

D i s t r i b u t i o n: Taiwan (Taikokan, terra typica).

## Genus Maikona MATSUMURA

(fig. 4)

1928 *Maikona* MATSUMURA, Insecta matsum., Sapporo, **2**:127. – S p e c i e s t y p i c a: *Maikona jezoensis* MATSUMURA (by monotypy).

D i a g n o s i s: Palpi upturned, with the first and the second joints very hairy, and the third joint shorter than the second, slender, naked, porrect. Antennae of the male (broken off), of the female setiform, slenderer towards the extremity. Frons with a beak-like triangular prominence. Thorax and legs with long pilosity. Base of abdomen with a large upright crest; remaining segments with very small crests. Nervation: fore-wings with vein 2 from four-fifths of the cell; 3, 4, 5 separated, distance between 3 and 4 greater distally than distance between 4 and 5; discocellulars incurved; 6 from the upper cell-angle; areole very small and narrow; 7 very long-stalked with 8 + 9 and 10, from the extremity of the areole. Hind wings with vein 2 from two-thirds of cell; 3, 4 and 6, 7 from the cell-angles; discocellulars incurved; 5 rather weak, from middle of discocellulars.

Male genitalia: Uncus relatively small, compressed, well arched with a small terminal hook. Tubus analis short, linguiform. Valva elongate and broad, more or less C-shaped, with the termen broad and nearly straight; fold of the sacculus rather narrow, bearing subterminally a small nearly straight, rather broad harpe, and ending in a narrow process, protruding beyond the termen. Aedeagus slightly shorter than the fold of the sacculus, moderately robust, nearly straight, bearing subterminally a dorsal and two lateral patches of minute cornuti; vesica relatively very large; fultura inferior roughly quadrate, with the proximal margin excised at middle, the sides produced and pointed at the proximal angle and slightly diverging distally, the distal margin very faintly incurved; fultura superior membraneous. Saccus short, rounded, with the proximal margin somewhat flattened.

D i s t r i b u t i o n: Japan.

### Maikona jezoensis (MATSUMURA)

1928 *Maikona jezoensis* MATSUMURA, Insecta Matsum., Sapporo, **2**:127, text-fig. (♀). – T e r r a t y p i c a: Hokkaido, Japan (Type, ♀, in Museum Hokkaido University). Neotypus, T e r r a t y p i c a: Josankei, Sapporo, Hokkaido, Japan (Type, ♂, in British Museum).

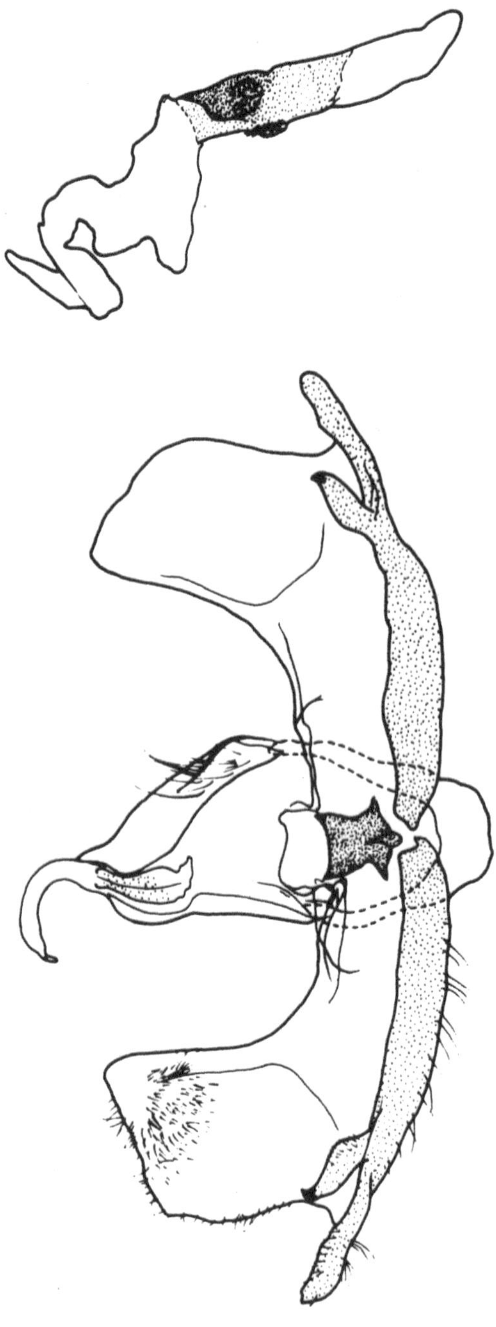

Fig. 4. *Maikona jezoensis* MATSUMURA. Neotype. B. M. Noct. 8504

D i a g n o s i s : Male: Palpi and head black; periocular region with some yellowish hair; base of the collar yellowish white, rest black; thorax black above; tegulae edged with yellowish white; underside and pilosity of the legs mixed with black and yellowish white; pilosity at the base of the abdomen yellowish white; the large crest deep black; pilosity of the abdomen black above, with fine yellowish white rings to the segments; becoming broader laterally; underside mixed with yellowish and dark grey; anal tuft black tipped with yellowish. Fore wings black; base of the dorsum pale orange; a fine rusty red streak from the base below the cell, curving towards the tornus and reaching the subterminal line; cell- and discocellular-marks dark purplish-grey edged with black, the former rounded, the latter fabiform; before the cell-spot and between the marks some rusty-red; veins finely yellowish white in the dorsal half of the wing; a rather ill-defined yellowish white spot below the base of vein 2; a large silky white subapical patch, faintly tinged with yellow, from the costa to vein 4, vaguely semilunar and edged with deep black, with its proximal black edge occupying the whole space between the discocellular mark and the patch; outer line double, fine, purplish-grey, following the distal margin of the subapical patch, then running obliquely inwards to below vein 2; terminal line rusty red, preceded by some black spots and edged distally with pale yellow; cilia blackish tipped with white. Hind wings pale yellow with a round grey-brown discocellular spot and a broad grey-brown terminal band, about 4 mm broad, hardly narrowing towards the anal angle, and with its proximal outline incurved and rather irregular; terminal line very fine, rusty red; cilia yellowish, whiter terminally, with an incomplete dark grey medial line; pilosity of the anal angle and the margin pale yellow. Underside of the fore wings dark grey-brown with the base and the dorsal area pale yellow; veins distinctly pale yellow; subapical patch yellowish. Underside of the hind wings as the upperside. Length of the forewing 18 mm.

MATSUMURA described the female only (Hokkaido, IV–V, holotype and four females paratypes). Since I am not aware of a subsequent description of the male, the specimen described above will serve as the neotype: male, Jozankei, Sapporo, Hokkaido, Japan, 18. IV. 1965, in the British Museum (Natural History).

Female: only slightly differing from the male. Tegulae without pale edges; abdomen above dark grey, laterally yellowish. Fore wings with the subapical blotch larger, broadening below vein 6, so that it is roughly rounded with a narrow costal portion. Hind wings with the dark terminal band somewhat broader (5 mm), and with some dark scaling between the band and the discocellular mark. Length of the fore wing 19 mm.

D i s t r i b u t i o n : Japan (Hokkaido = terra typica).

## Genus **Exsula** JORDAN

(fig. 5)

1896 *Exsula* JORDAN, Novit. Zool., 3:35, 41. – S p e c i e s t y p i c a : *Eusemia dentatrix* WESTWOOD (by original designation).

D i a g n o s i s : Proboscis fully developed; antennae slightly dilated towards the extremity; palpi upturned; 1st and 2nd joints of palpi with long but rather sparse hair in front; the 3rd well developed, slightly clubbed, nearly naked and somewhat porrect; frons convex with a rather weak, slightly rounded prominence; eyes naked; tibiae and tarsi with appressed scales. Venation: in the fore wings vein 2 from about 3/4 of cell; 3, 4 and 5 rather approximated, nearly equidistant; 6 from lower cell-angle; areole very narrow; 7 and 10, 8 + 9 from end of areole; 11 from cell. In the hind wings, vein 2 from 2/3 of cell; 3, 4 and 6, 7 from a point; discocellulars weakly incurved; 5 obsolescent, from middle of discocullulars; 8 as in the related genera.

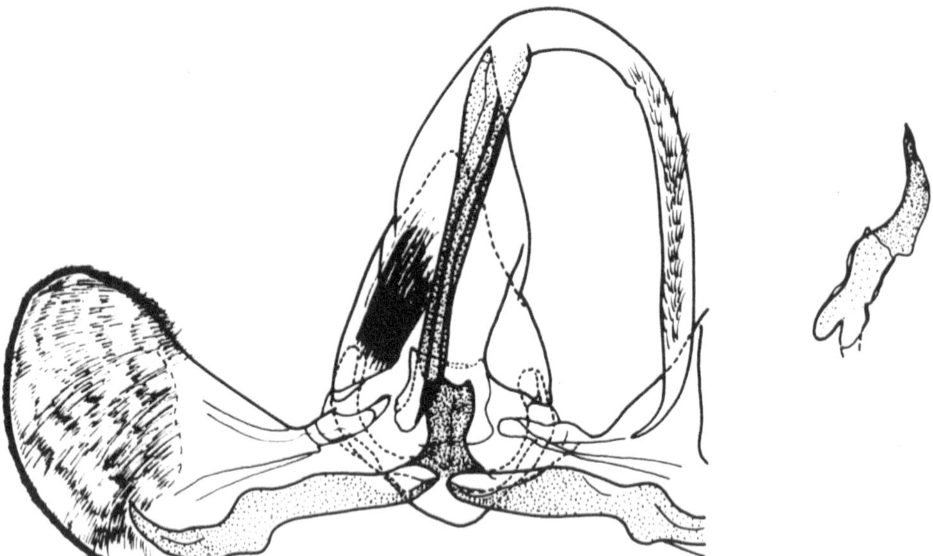

Fig. 5. *Exsula dentatrix* (WESTWOOD). B. M. Noct. 8436

Male genitalia. Uncus very long, longer than costa, compressed, strongly arched at base, then straight, with a terminal hook. Tubus analis nearly as long as the uncus, sclerified. Valva roughly pear-shaped, narrow basally, oval in the distal 3/4; fold of sacculus rather narrow, ending in a short, sickle-shaped harpe. Aedeagus very short, but slightly longer than the harpe, relatively robust, sickle-shaped in the distal half; fultura inferior broad at base, then narrowing, and again broadening, with the terminal margin excised. Saccus hardly developed, triangular.

D i s t r i b u t i o n: China, North India, Burma.

Key to the species

1 Hind wing orange, with black discocellular patch and terminal band . . . *E. dentatrix*, p. 31
1′ Hind wing black, with a blue terminal band . . . . . . . . . . . . . . . . . . . *E. victrix*, p. 32

## Exsula dentatrix (WESTWOOD)

1848 *Eusemia dentatrix* WESTWOOD, Cat. Or. Entomol.: 68; pl. 33, fig. 5 (♀). – Terra typica: Khasis, Assam (Type, ♀, Museum Oxford University).
1892 *Episteme dentatrix* – KIRBY, Syn. Cat. Lepidopt. Het.: 27.
1896 *Exsula dentatrix* – JORDAN, Novit. Zool., 3:35.

Diagnosis: Head, thorax and fore wings black; 1st and 2nd joints of palpi orange laterally; a postcephalic orange ring; fore coxae and pilosity of femora orange; abdomen orange banded with black, sometimes nearly black; base and anal tuft black in male, orange in female. Fore wings with orange or white markings; a square medial patch in cell, conjoined to a similar oblique patch below base of vein 2, sometimes forming a band; spots beyond the cell in interspaces III, IV and VI; metallic blue spots at base below costa and above vein 1; some blue on the disco-cellulars; veins of the terminal area streaked with blue. Hind wings orange yellow or orange red; costal area black, emitting conjoined spots at both cell-angles, giving off a short streak on distal side in discal fold; a broad terminal band black, veins on it streaked with blue. Length of fore wing 35 mm.

Early stages: Larva. First four rings black, the six following greyer, with narrow white bands, and white dorsal, subdorsal and lateral stripes. Head, pronotum, spots above prolegs and terminal segment light leather-brown with black dots and streaklets. Each segment ringed with 10 black warts bearing a long white seta. On *Saurauja tristyla*.

Distribution: North India to China (Taiwan).

### Exsula dentatrix dentatrix (WESTWOOD)

1912 *Exsula dentatrix dentatrix* – JORDAN in SEITZ, Großschmetterl. Erde, **11**:8; pl. 1e.
1912 *Exsula dentatrix dentatrix* forma *dentatrix* vera JORDAN in SEITZ, Großschmetterl. Erde, **11**:8.
1912 *Exsula dentatrix dentatrix* forma *conjuncta* JORDAN in SEITZ, Großschmetterl. Erde, **11**:8.

Distribution: Nepal, Sikkim, Assam.

### Exsula dentatrix nigridorsa (CANDEZE)

1927 *Exsula dentatrix* var. *nigridorsa* CANDEZE, Encycl. Entomol., B. **3**: Lepidopt., **2**:78. – Terra typica: North Vietnam (Type in Paris Museum).

Diagnosis: Abdomen black, with only 3 (male) or 4 (female) last segments banded with orange; subbasal spot in the fore wings absent; 4 small white postmedian spots. Hind wings orange red with the terminal band very broad (1/4 of the wing-breadth), with a strong metallic blue gloss.

Distribution: Cambodia, North Vietnam, Kwangtung.

### Exsula dentatrix albomaculata (MIYAKE)

1907 *Exsula dentatrix* var. *albomaculata* MIYAKE, Annot. zool. japon., Tokyo, 6:79. – Terra typica: Taiwan.
1936 *Exsula dentatrix albomaculata* forma *flavomaculata* MELL, Stettin. Entomol. Ztg., 97: 167.

Diagnosis: Fore wings with the markings white; a white subbasal spot. Hind wings with the orange colour reaching the costa.

Distribution: Taiwan (Terra typica), Hainan, Kwangsi, Kwangtung coasts.

## Exsula victrix (WESTWOOD)

1848 *Eusemia victrix* WESTWOOD, Cab. Or. Entomol.: pl. 33, fig. 3 (♀). – Terra typica: Nepal (Type: ♀ in Mus. Oxford Univ.).
1892 *Episteme victrix* – KIRBY, Syn. Cat. Lepidopt. Het.: 26.
1936 *Exsula dentatrix victrix* – MELL, Stettin. Entomol. Ztg., 97:168.

Diagnosis: Body and wings black; palpi with some orange hair; a postcephalic orange ring; fore coxae and femora with orange pilosity; abdomen with blue reflections; lateral tufts of hair fulvous; extremity above and terminal half on underside orange. Fore wings with a pale yellow square patch in middle of cell extending almost to the costa and conjoined to a similar patch below the base of vein 1; a series of small white postmedial spots, sometimes absent; silvery blue subbasal spots below costa and cell; silvery blue points at cell-angles; veins in terminal area streaked with brilliant blue, that colour expanding towards the termen and confluent towards the apex. Hind wings with blue reflections; a brilliant blue band in terminal area, not reching the termen and varying in breadth. Length of fore wing 30 – 37 mm.

Distribution: Western Himalayas; Nepal, Burma to Yunnan.

### Exsula victrix victrix (WESTWOOD)

1875 *Eusemia silhetensis* BUTLER, Ann. Mag. Nat. Hist., London, (4) **15**:139 (♀). – Terra typica: Khasis, Assam (Types, ♂, ♀, in British Museum).
1875 *Eusemia tyrianthina* BUTLER, Entomologist's mon. Mag., **12**:116 (♂, ♀). – Terra typica: Sikkim (Types, ♂, ♀, in British Museum).
1912 *Exsula victrix victrix* – JORDAN in SEITZ, Großschmetterl. Erde, **11**:9; pl. 1d.
1912 *Exsula victrix* var. *burmaensis* STRAND, Lepidopt. Cat., **5**:7. – Terra typica: Khyen Hills, Burma (Types, ♂, ♀, in British Museum).
1912 *Exsula victrix victrix* forma *victrix* – JORDAN in SEITZ, Großschmetterl. Erde, **11**:9.
1912 *Exsula victrix victrix* forma *punctigera* JORDAN in SEITZ, Großschmetterl. Erde, **11**:9.
1912 *Exsula victrix victrix* forma *limma* JORDAN in SEITZ, Großschmetterl. Erde, **11**:9. – Terra typica: Not stated.

Distribution: Nepal and Burma to Vietnam and Yunnan.

## *Exsula victrix orientalis* (BUTLER)

1875 *Eusemia orientalis* BUTLER, Ann. Mag. Nat. Hist., London, (4) **15**:139 (♀). – Terra typica: Mussorie, N. W. Himalaya (Type, ♀, in British Museum).
1901 *Exsula victrix orientalis* – HAMPSON, Cat. Lepidopt. Phal. Brit. Mus., **3**:532.

Diagnosis: Fore wings with the postdiscal spots much approximated to each other, sometimes with an additional white spot in interspace. Hind wings with the blue band very narrow.

Distribution: West Pakistan (Mussoree, Garwhal).

## Genus **Philippodamias** CLENCH

(fig. 6)

1958 *Philippodamias* CLENCH, Ann. Carnegie Museum, Pittsburgh, **35** (6): 73. – Species typica: *Philippodamias jocelyna* CLENCH (by original designation).

Diagnosis: Antennae thickened before the extremity; frons with a slight prominence; 1st and 2nd joints of palpi fringed with long hair below, upturned; 3rd joint short, about half 2nd, covered with appressed scales, pointed, porrect; eyes naked; proboscis fully developed; mid and hind femora fringed with sparse long hair; mid and hind tibiae with appressed scales and hairs; hind tibiae bearing two pairs spurs; tarsal claws of all legs with a small inner tooth. Nervation: fore wing: vein 2 from 4/5 of cell; 3, 4, 5 separated, about equidistant; upper discocellular weakly angled at middle; 6 from the upper cell-angle; an areole present; 7 from extremity of the areole, slightly separated from 8, 9 and 10 which are long-stalked. Hind wing: vein 2 from beyond 4/5 of cell; 3, 4 much approximated; 5 from middle of the discocellulars; 6, 7 from angle.

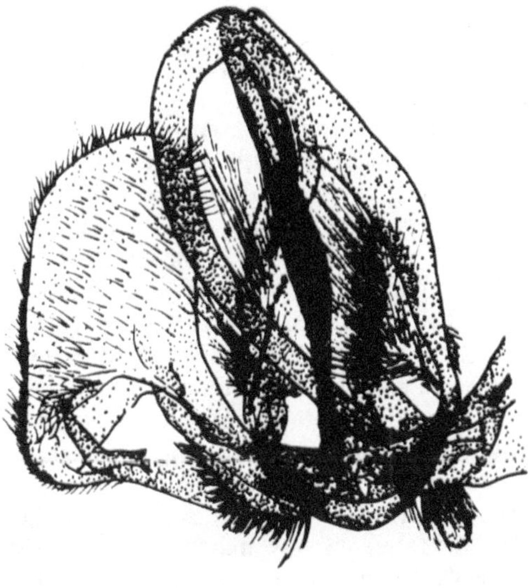

Fig. 6. *Philippodamias jocelyna* CLENCH. Type. Carnegie Mus., slide C-1213

Male genitalia: Uncus very long, slightly longer than the valva, slender, arched in the basal half, then nearly straight, with a small terminal hook. Vincula moderately broad, fringed with stiff hair. Valva narrow at base, then broadening and roughly oval, with the apex broadly rounded and the sacculus distinctly angled at middle; harpe placed at middle of the basal straight portion of sacculus, short, hooked. Aedeagus short, nearly straight, slightly more robust distally; fultura inferior little differentiated. Saccus very short, rounded.

Distribution: Philippine Islands.

## **Philippodamias jocelyna** CLENCH

1958 *Philippodamias jocelyna* CLENCH, Ann. Carnegie Mus., **35**(6):73 (♂). – Philippine Islands (Type, ♂, in Carnegie Museum).

Diagnosis: Body and appendages bluish or greenish black. Fore wings bluish black; a broad nearly straight yellowish orange band from middle of costa to near tornus, ending rounded just before tornus. Hind wings black, tinged with brown distally, with bluish-purple iridescence, especially in the basal area; lower angle of cell bearing a small orange patch. Cilia of both wings concolorous. Length of fore wing 29,5 mm.

Distribution: Philippine Islands: Matuginao, Samar Island (terra typica).

## Genus **Scrobigera** JORDAN

(fig. 7)

1896 *Scrobigera* JORDAN, Novit. Zool., **3**:37, 40. – Species typica: *Eusemia amatrix* WESTWOOD (by original designation).

Diagnosis: Proboscis fully developed; palpi upturned, the 2nd joint fringed with long hair in front, the 3rd long, prismatic, porrect, longer in female; frons with very indistinct rounded prominence; antennae slightly dilated towards extremity; tibiae and tarsi with appressed scales; hind tibiae with two pairs of spurs. Venation: fore wings with vein 2 from just beyond middle of cell; distance between 3 and 4 slightly greater than distance between 4 and 5; discocellulars incurved; areole small, narrow; 6 from angle of cell; 7 and 10, (8 + 9) from extremity of areole. Male with a fringe of long black hair on upperside in discal fold, in and beyond end of cell, in hind wing, so that the median fold is strongly protruding beneath; vein 2 from 4/5 of cell; 3, 4 and 6, 7 from cell-angles; discocellulars angled outwards at the origin of 5 which is obsolescent and given off from middle of discocellulars.

Male genitalia: Uncus very long, distinctly longer than costa of valva, compressed, arched except in terminal third, and ending in a small hook; tubus analis as long as the uncus, slender, sclerified. Vincula with a strong bunch of stiff hair. Valva roughly pear-shaped, narrow in the proximal 2/5, with the costa faintly concave, the apex rounded, the termen evenly arched, the sacculus straight in the proximal half, then broadly outcurved, bearing at the end of the narrow fold a rather short, sickle-shaped harpe, with a subbasal knob on the costal margin.

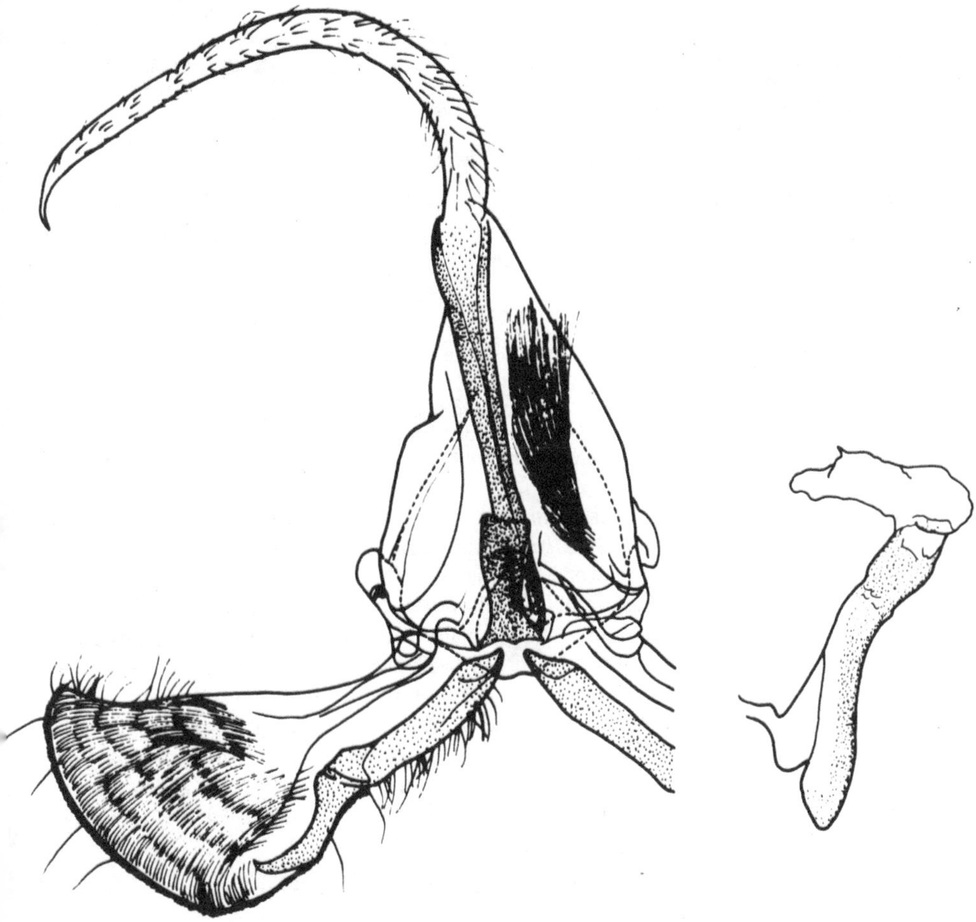

Fig. 7. *Scrobigera amatrix* (WESTWOOD).

Aedeagus somewhat shorter than costa, moderately robust, slightly arched; fultura inferior broad, produced and rounded medially with the lateral margins concave. Saccus hardly differentiated, rounded.

D i s t r i b u t i o n: China, Indo-China, North India, Indonesia.

Key to the species

1 Hind wings with an orange band at least in male . . . . . . . . . . . . . . . . . . . . 2
1' Hind wing without orange band . . . . . . . . . . . . . . . . . . . . . . . . . . . 4
2 Female also with an orange band on hind wing . . . . . . . . . . . . . . . . . . . . 3
2' Female without an orange band on hind wing . . . . . . . . . . . **S. vacillans**, p. 38
3 Fore wing with 4 yellowish spots beyond cell . . . . . . . . . . . **S. amatrix**, p. 36
3' Fore wing with a yellowish bar beyond cell . . . . . . . . . . . . **S. proxima**, p. 37
4 Fore wing unspotted . . . . . . . . . . . . . . . . . . . . . . . . . . . . . . . 5
4' Fore wing with orange, yellow or white spots or band . . . . . . . . . . . . . . . . 7

5 Costa of forewing yellow . . . . . . . . . . . . . . . . . . . . . . . . . . . *S. umbrosa*, p. 41
5′ Costa of forewing black . . . . . . . . . . . . . . . . . . . . . . . . . . . . . . . . . . . . . . . . . 6
6 Cilia of both wings white . . . . . . . . . . . . . . . . . . . . . . . . *S. albomarginata*, p. 39
6′ Cilia only partly pure white . . . . . . . . . . . . . . . . . . . . . . . . . . . . . . *S. vitalisi*, p. 42
7 Fore wing with the medial series of yellowish spots and four spots beyond cell. . . . . . *S. vulcania*, p. 40
7′ Fore wing with an oblique orange, yellow or white band . . . . . . . . . . . . . . . . . 8
8 Fore wing with an oblique orange band . . . . . . . . . . . . . . . . . . . . *S. semperi*, p. 39
8′ Fore wing with the oblique band pale yellow or white . . . . . . . . . . . . . . . . . . . . 9
9 Fore wing with an oblique white band; cilia white at apex . . . . . . *S. niveifasciata*, p. 39
9′ Fore wing with an oblique pale yellow band; cilia concolorous, without white . . . . . . *S. claggi*, p. 41

## Scrobigera amatrix (WESTWOOD)

1848 *Eusemia amatrix* WESTWOOD, Cab. Or. Entomol.: 68; pl. 33, fig. 4. – Terra typica: Sikkim (Type, ♂, in Museum Oxford University).
1896 *Scrobigera amatrix* – JORDAN, Novit. Zool., 3:40.

Diagnosis: Head, thorax and wings black; palpi, frons and collar orange mixed with blackish in male, black in female; occiput with a white line in female; pectus with orange pilosity; abdomen orange with base and transverse lines black; anal tuft mixed with black; in female black more extensive. Fore wings with blue reflections; base of costa orange in male; a pale yellow square spot in cell end; a small elongate spot at base of vein 2 and an oval spot below it; a series of four smaller postmedial spots in intervals IV–VII; basal area with silvery-blue streaks; a blue oblique bar across the cell, and another on the discocellulars; cilia concolorous, but white apically. Hind wings with a broad, waved orange band with a black dot on it at the upper cell-angle; a tuft of long black hair on discal fold; cilia concolorous, but white apically in male, tipped with white in female. Length of fore wing 33 – 38 mm.

Distribution: Sikkim, Bhutan, Assam, Vietnam, West China.

### Scrobigera amatrix amatrix (WESTWOOD)

1859 *Eusemia aruna* MOORE, Lepidopt. East Ind. Co.: 288 (♂, ♀). – Terra typica: Sikkim (Type in British Museum).
1912 *Scrobigera amatrix amatrix* – JORDAN in SEITZ, Großschmetterl. Erde, 11:7; pl. 2c.

Distribution: Assam, Sikkim.

### Scrobigera amatrix media JORDAN

1912 *Scrobigera amatrix media* JORDAN in SEITZ, Großschmetterl. Erde, 11:7 (♂). – Terra typica: Chien Hoa, North Vietnam (Type, ♂, in British Museum).

Diagnosis: Seventh urite nearly wholly black; preceding urites with yellow bands; fore wings with the orange band narrower; cellular spot connected with the basal area; cilia white.

Distribution: Central Vietnam.

### Scrobigera amatrix chinensis JORDAN

1912 *Scrobigera amatrix chinensis* JORDAN in SEITZ, Großschmetterl. Erde, **3**:6; pl. 1c. – Terra typica: Chekiang, West China (Type, ♂, in British Museum).

Diagnosis: Abdomen black dorsally, with orange bands; anal tuft black tipped with orange; fore wings with the postmedial spots sometimes white.

Distribution: West and Central West China (Chekiang, North Kwangtung).

## Scrobigera proxima (WALKER)

1854 *Eusemia proxima* WALKER, List Lepidopt. Het. Brit. Mus., **1**:50 (♂). – Terra typica: Assam (Type, ♂, in British Museum).
1896 *Scrobigera proxima* – JORDAN, Novit. Zool., **3**:37.

Diagnosis: Head, thorax and wings black; 1st and 2nd joints of palpi, sides of frons, collar, legs and abdomen orange, the latter with the base and the two last segments black, and sometimes with obsolescent dark dorsal spots; in female, frons and collar black, and a pale yellow neck-ring. Fore wings with a medial pale yellow oblique band of three confluent spots, sometimes interrupted, from subcosta in male, from costa in female; a postdiscal narrow band between veins 3 and 7, sometimes absent; an orange costal streak; silvery-blue spots in middle of cell and on discocellulars, cilia white apically. Hind wings with a medial orange or orange-red band, from cell to anal margin, in female its distal edge sinuous and angled in discal fold; latter blackish in male; cilia white apically. Length of fore wing 28 – 31 mm.

Early stages: Larva black; head, pronotum, pectus, legs and terminal segment yellow with sparse black hair; a dorsal and a lateral stripe on each side white; intersegmental bands pink and white. On *Cissus*.

Distribution: Assam to Indonesia.

### Scrobigera proxima proxima (WALKER)

1848 *Eusemia amatrix* – WESTWOOD (partim), Cat. Entomol. Or.: 68.
1874 *Eusemia clymene* BOISDUVAL, Rev. Mag. Zool., (3) **2**:97. – Terra typica: Assam (Type in British Museum).
1892 *Episteme clymene* – KIRBY, Syn. Cat. Lepidopt. Het.: 28.
1901 *Eusemia proxima* subsp. *clymene* – HAMPSON, Cat. Lepidopt. Phal. Brit. Mus., **3**:541.
1912 *Scrobigera proxima proxima* – JORDAN in SEITZ, Großschmetterl. Erde, **11**:7.

Distribution: Assam (Terra typica), Burma, Malaya, Sumatra, Java, Batoe Islands.

### Scrobigera proxima hesperioides (WALKER)

1862 *Eusemia hesperioides* WALKER, J. Linn. Soc. London, **6**:86 (♀). – Terra typica: Borneo (Type, ♂, in Museum Oxford University).
1875 *Eusemia tricolor* BUTLER, Ann. Mag. Nat. Hist., London, (4) **15**:142 (♀). – Terra typica: Kina Baloe, Borneo (Type, ♀, in British Museum).

1875 *Eusemia pulchra* BUTLER, Ann. Mag. Nat. Hist., London, (4) **15**:143; pl. 13, fig. 4 (♂, ♀). – Terra typica: "Murrut, India" (ex errore) = Sarawak, Borneo (HAMPSON 1901) (Type, ♂, in British Museum).
1892 *Episteme hesperioides* – KIRBY, Syn. Lepidopt. Het.: 28.
1892 *Episteme tricolor* – KIRBY, Syn. Cat. Lepidopt. Het.: 28 (♀).
1892 *Episteme pulchra* – KIRBY, Syn. Cat. Lepidopt. Het.: 28.
1894 *Episteme staudingeri* OBERTHÜR, Etudes entomologiques, **19**:22; pl. 3, fig. 15 (♂). – Terra typica: Borneo (Type, ♂, in British Museum).

Diagnosis: Fore wings with the discal band absent or at most represented by a few yellow scales; hind wings with the discal band variable, sometimes only a few mm broad, sometimes much broader.

Distribution: Borneo.

## Scrobigera vacillans (WALKER)

1864 *Eusemia vacillans* WALKER, List Lepidopt. Het. Brit. Mus., **31**:51 (♀). – Terra typica: Celebes (Type, ♀, in Museum Oxford University).
1892 *Episteme vacillans* – KIRBY, Syn. Cat. Lepidopt. Het.: 28.
1902 *Scrobigera vacillans* – JORDAN in SEITZ, Großschmetterl. Erde, **11**:7.

Diagnosis: Head, thorax and wings black; 1st and 2nd joints of palpi orange; frons and collar orange in male, black in female; latter with a white ring around the neck; legs orange; abdomen orange with black dorsal bands in male, black in female. Fore wings with a costal basal orange streak in male; an oblique pale yellow bar across end of cell; a small spot above base of vein 2; a large spot below it; an oblique bar beyond cell in interspaces III–VI, with dentate distal edge; silvery-blue streaks below costa, on vein 1, and an oblique bar in middle of cell with another on the discocellulars; cilia white apically. Hind wings with a diffuse light orange medial band in male, expanding in discal fold and on anal margin; cilia white at tips in male, white in female. Length of fore wing 28 – 30 mm.

Distribution: Celebes.

### Scrobigera vacillans vacillans (WALKER)

1875 *Eusemia eudamioides* BUTLER, Entomologist's mon. Mag., **12**:124 (♀). – Terra typica: Celebes (Type, ♀, in British Museum).
1892 *Episteme eudamioides* – KIRBY, Syn. Cat. Lepidopt. Het.: 28.
1912 *Scrobigera vacillans vacillans* – JORDAN in SEITZ, Großschmetterl. Erde, **11**:7.

Distribution: South Celebes.

### Scrobigera vacillans taeniata ROTHSCHILD & JORDAN

1903 *Scrobigera vacillans* ssp. *taeniata* ROTHSCHILD & JORDAN, Novit. Zool., **10**:487 (♂, ♀). – Terra typica: Sawagan, North Celebes (Types, ♂, ♀, in British Museum).

D i a g n o s i s : Fore wings in male with the two lower spots of the medial series confluent and almost conjoined to the upper bar-shaped spot; in female, markings whiter and postmedial bar narrower; hind wings with cilia as in male, not wholly white.

D i s t r i b u t i o n : North Celebes.

## Scrobigera niveifasciata ROTHSCHILD

1896 *Scrobigera niveifasciata* ROTHSCHILD, Novit. Zool. **3**:38 (♀). – T e r r a t y p i c a : Borneo (Type, ♀, in British Museum).

D i a g n o s i s : Body and wings black; antennae and sides of frons with some white scales; first and second joints of palpi orange; neck with a white ring; legs and underside of abdomen orange; upperside of tibiae and tarsi with some brown sprinkling. Fore wing with an oblique white band from middle of costa to just above tornus; no other markings, except some blue scales in middle of cell and on the discocellulars; cilia white apically. Hind wings unmarked; cilia with some white hairscales and white apically. Length of fore wing 32 mm.

D i s t r i b u t i o n : Borneo.

## Scrobigera semperi (FELDER)

1868–1874 *Agarista semperi* FELDER, Reise Novara: pl. 107, fig. 4 (♀). – T e r r a t y p i c a : Philippine Islands (Type, ♀, in British Museum).
1874 *Eusemia semperi* – BOISDUVAL, Rev. Mag. Zool., (3) **2**:101.
1892 *Episteme semperi* – KIRBY, Syn. Cat. Lepidopt. Het.:28.
1892 *Eusemia milioniata* SWINHOE, Cat. Het. Mus. Oxon.:162; pl. 5, fig. 2 (♂). – T e r r a t y p i c a : Philippine Islands (Type, ♂, in Museum Oxford University).
1896 *Scrobigera semperi* – JORDAN, Novit. Zool. **3**:37.

D i a g n o s i s : Body black with faint blue reflections; basal half of palpi orange; fore coxae, fore femora and tibiae orange below. Fore wings bluish black; basal part of costa orange in male; from middle of costa to near tornus an oblique orange band with irregular edges; a faint subcostal silvery-blue streak; a silvery-blue bar across middle of cell and another on the discocellulars; cilia white apically. Hind wings black with stronger blue reflections, unmarked; cilia wholly white. Length of fore wing 26 – 31 mm.

D i s t r i b u t i o n : Philippine islands (Luzon, Mindanao).

## Scobigera albomarginata (MOORE)

1872 *Eusemia albomarginata* MOORE, Proc. Zool. Soc. London, **1872**:509 (♂). – T e r r a t y p i c a : Burma (Type, ♂, in British Museum).
1892 *Agarista albomarginata* – KIRBY, Syn. Cat. Lepidopt. Het.:21.
1896 *Scrobigera albomarginata* – JORDAN, Novit. Zool. **3**:37.

Diagnosis: Body black; 1st and 2nd joints of palpi and sides of frons mixed with orange in male, with yellowish on palpi and white on frons in female; fore and mid femora and tibiae and parts of hind femora and tibiae orange; abdomen with blue reflections, banded with orange on the underside. Fore wings black with blue reflections; a basal costal orange streak in male; costa, media and vein 1 streaked with metallic blue; an oblique bar across middle of cell and a discocellular bar also metallic blue; other markings small, white, often absent on the upperside, especially in male; when present, consisting of a tripartite cell-mark, a spot above cell, and a series of 3–4 postmedial dots. Hind wings with a stronger blue sheen, unspotted. In both wings, cilia white, often expanding towards apex, sometimes on termen; sometimes cilia yellowish. Length of fore wing 27 – 40 mm.

Distribution: Burma, Vietnam, Tenasserim, Andaman Islands, Philippine Islands.

### Scrobigera albomarginata albomarginata (MOORE)

1885 *Eusemia opheltes* DRUCE, Proc. Zool. Soc. London, **1885**:518. – Terra typica: Burma (Type, ♂, in British Museum).
1891 *Eusemia candidemarginata* POUJADE, La Nature, **13**:143; text-fig. – Terra typica: Burma.
1892 *Agarista opheltes* – KIRBY, Syn. Cat. Lepidopt. Het.:21.
1912 *Scrobigera albomarginata albomarginata* – JORDAN in SEITZ, Großschmetterl. Erde, **11**:8; pl. 2c.

Diagnosis: As mentioned above for the species. Cilia white.

Distribution: Burma, Tenasserim, Vietnam.

### Scrobigera albomarginata flaviciliata (BOISDUVAL)

1874 *Agarista flaviciliata* BOISDUVAL, Rev. Mag. Zool., (3) **2**:104 (♀). – Terra typica: Philippine Islands (Type, ♀, in British Museum).
1874 *Agarista fimbriata* BOISDUVAL, Rev. Mag. Zool., (3) **2**:pl. 8, fig. 1. – Terra typica: Philippine Islands (Type, ♂, in British Museum).
1901 *Scrobigera albomarginata* subsp. *flaviciliata* –HAMPSON, Cat. Lepidopt. Phal. Brit. Mus., **3**:544.

Diagnosis: Cilia yellowish in both wings.

Distribution: Philippine Islands (Luzon).

### Scrobigera vulcania (BUTLER)

1875 *Eusemia vulcania* BUTLER Entomologist's mon. Mag., **12**:123 (♂). – Terra typica: Burma (Type, ♂, in British Museum).
1881 *Eusemia glossatrix* WESTWOOD, Oates' Matabeleland:356 (♂). – Terra typica: "South Africa" (ex errore) = Burma (ROTHSCHILD 1899) (Type in British Museum).
1892 *Episteme vulcania* – KIRBY, Syn. Cat. Lepidopt. Het.:28.
1892 *Xanthospiloteryx glossatrix* – KIRBY, Syn. Cat. Lepidopt. Het.:29.
1896 *Scrobigera vulcania* – JORDAN, Novit. Zool., **3**:37.
1912 *Scrobigera vulcanica* –STRAND, Lepidopt. Cat., **5**:10.

D i a g n o s i s : Body black; palpi, sides of frons and collar marked with orange; in female, collar unmarked and neck with a white ring; femora and tibia mostly orange; abdomen banded with orange on underside, but sometimes unbanded; extremity orange, sometimes on underside only. Fore wings black with blue reflections; costal streak orange at base in male; a pale yellow spot in middle, composed of a square spot in cell-tip and spots above and below the base of vein 2; a postdiscal series of four white spots in interspaces V–VIII; cilia concolorous, white apically. Hind wings black with strong blue reflections; cilia white. Length of forewing 29 – 30 mm.

D i s t r i b u t i o n : Burma, Tenasserim, Vietnam.

### Scrobigera vulcania vulcania (BUTLER)

D i a g n o s i s : Extremity of abdomen orange above and below in female.

D i s t r i b u t i o n : Burma and Tenasserim.

### Scrobigera vulcania melaina (JOANNIS)

1900 *Scrobigera melaina* JOANNIS, Bull. Soc. entomol. France, **1900**:280. – T e r r a t y p i c a : Kaobang, Indo-China (Type, ♀, in Museum Paris).
1936 *Scrobigera vulcanica melaina* – JORDAN in SEITZ, Großschmetterl. Erde, **11**:8 (♀).

D i a g n o s i s : Extremity of abdomen wholly bluish black in female, as the rest of abdomen, with only the extreme tip orange on the underside. Cellspot on fore wing larger than the third spot of the median band; white on cilia not expanding onto wing surface.

D i s t r i b u t i o n : Vietnam (Caobang = Terra typica).

### Scrobigera claggi CLENCH

1953 *Scrobigera claggi* CLENCH, Ann. Carnegie Mus., **33** (6):141. – T e r r a t y p i c a : Philippine Islands (Type, ♂, in Carnegie Museum).

D i a g n o s i s : Differs from *Scrobigera niveifasciata* ROTHSCHILD in the yellow, not white median band on fore wings and in the absence of white on the cilia. – According to CLENCH (1953), doubtfully a true *Scrobigera*, possibly belonging in the genus *Philippodamias* CLENCH.

D i s t r i b u t i o n : Philippine Islands.

### Scrobigera umbrosa CLENCH

1953 *Scrobigera umbrosa* CLENCH, Ann. Carnegie Mus., **33** (6):141. – T e r r a t y p i c a : Manila, Philippine Islands (Type, ♂, in Carnegie Museum).

D i a g n o s i s : Differs from *Scrobigera semperi* FELDER as follows: antennae reddish below (uniformly black in *S. semperi*); forepart of patagia yellowish, not black; sides of frons yellowish white, not white; fore wings without the yellowish band; costal streak yellow in male, not black; fore wings on underside with a narrower orange band, without the large black square spot of *S. semperi*, and becoming obsolescent towards the anal region (in *S. semperi* the band is even strong in the anal region as elsewhere).

D i s t r i b u t i o n : Philippine Islands (Manila = Terra typica).

### **Scrobigera vitalisi** CANDEZE

1927 *Scrobigera vitalisi* CANDEZE, Encycl. entomol., B. 3, Lépidopt. **2**:79 (♂, ♀). – T e r r a t y p i c a : Pnom-Penh, Vietnam (Types, ♂, ♀, in Paris Museum).

D i a g n o s i s : Body black; first and second joints of palpi and sides of frons mixed with orange; femora and tibiae orange; abdomen with orange bands below. Fore wings with a broad, rather indistinct spot at cell-middle; another smaller between veins 2 and 3, both of bluish-white scales. Fore wings with cilia white from apex to vein 7, rest concolorous in male, variegated with brown in female, where there is a broad apical zone from costa to vein 7 strongly sprinkled with white scales, except on veins. Hind wings with cilia white; on the anal angle a bluish black area; cilia preceded by a white terminal band with the veins concolorous, broader in female.

D i s t r i b u t i o n : Vietnam (Pnom-Penh).

## Genus **Fleta** JORDAN

(fig. 8)

1896 *Fleta* JORDAN, Novit. Zool., **3**:34. – S p e c i e s t y p i c a : *Agarista belangeri* GUÉRIN–MÉNÉVILLE (by original designation).

D i a g n o s i s : Proboscis fully developed; antennae moderately dilated towards the extremity; palpi upturned; 1st and 2nd joints fringed with long hair in front; 3rd moderate, smoothly scaled; frons with the prominence hardly developed; hind tibiae with sparse long hairs. Venation: Fore wings with vein 2 from very near cell-extremity; 3, 4 from a point; 5 from the lower quarter of discocellulars; latter faintly angled outwards at their origin; 6 from upper cell-angle; areole minute, sometimes reduced to a point; 7, 10, 8 + 9 from extremity of areole. Hind wings with vein 2 from 5/7 or 6/7 of cell; 3, 4 from a point; 5 obsolescent, from below upper angle of cell, sometimes from middle of discocellulars (*F. moorei*); 6, 7 very short stalked.

Male genitalia: Uncus slightly longer than sacculus, narrow, strongly curved in basal section, faintly curved in distal section, bearing a small terminal hook with a weak tooth. Vincula moderately broad, without projections, fringed with long, stiff hair. Valva banana-shaped, rather strongly curved, costa and sacculus nearly parallel, apex broadly rounded; basal fold of sacculus reaching costal area, seemingly

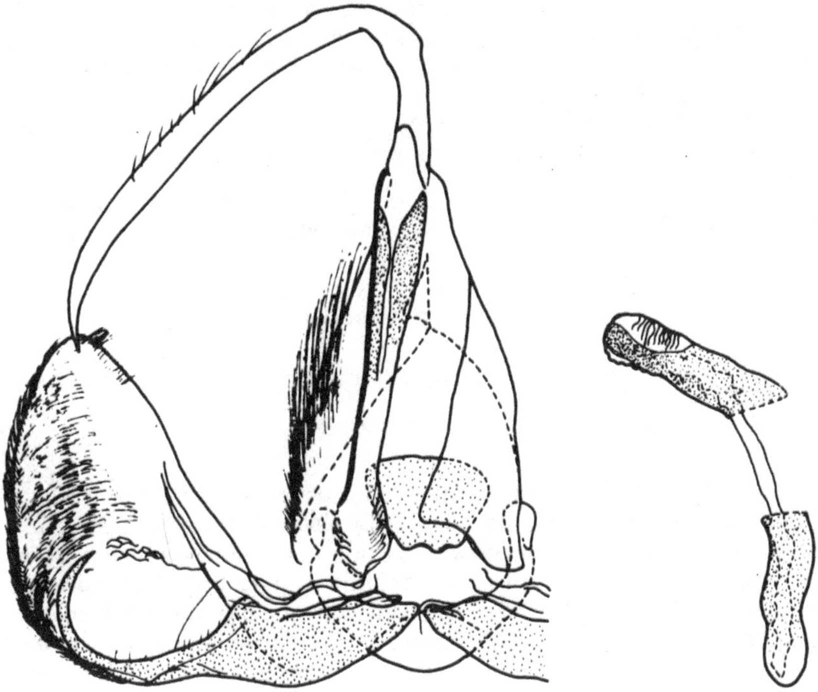

Fig. 8. *Fleta belangeri* (GUERIN). Type. B. M. Noct. 8505

fused with a costal rib, occupying about one fifth of the valva and ending in a slender, strongly arched harpe. Aedeagus relatively small, about one-third of valva in length, rather slender, regularly arched; fultura inferior narrow. Saccus very short, rounded.

Distribution: Java, Sumatra.

Key to the species

1 Hind wings with the black margin continuing along the anal margin . . . ***F. belangeri***, p. 43
1' Hind wing with the black margin narrowing towards the anal angle and not continuing along the anal margin . . . . . . . . . . . . . . . . . . . . . . . . . ***F. moorei***, p. 44

## **Fleta belangeri** (GUÉRIN–MÉNÉVILLE)

1834 *Agarista belangeri* GUÉRIN–MÉNÉVILLE, Voyage Bélanger aux Indes Orientales:506; pl. 5, fig. 3. – Terra typica: Java (Type, ♂, in British Museum).
1854 *Eusemia amatrix* var. ? ? WALKER, List Lepidopt. Het. Brit. Mus., 1:49. –
1874 *Eusemia belangeri* – BOISDUVAL, Rev. Mag., Zool., (3) 2:88.
1892 *Episteme belangeri* – KIRBY, Syn. Cat. Lepidopt. Het.:27.
1896 *Fleta belangeri* – JORDAN, Novit. Zool., 3:34.

Diagnosis: Head, body and appendages blackish brown; head and collar mixed with yellow; pilosity of pectus and legs yellow, the latter mixed with white; abdomen orange laterally, yellow ventrally in male, yellowish with the base blackish brown in female. Wings blackish brown. Fore wings with an oblique yellow band beyond cell, between veins 2 and 7. Hind wings with a broad median area orange-red; the black margin continued along the anal margin. Length of fore wing 20 – 23 mm.

Distribution: East and West Sumatra, West Java (in hilly country).

### Fleta moorei (FELDER)

1874 *Agarista moorei* FELDER, Reise Novara: pl. 107, fig. 5 (♂). – Terra typica: Java (Type, ♂, in British Museum).
1892 *Episteme moorei* – KIRBY, Syn. Cat. Lepidopt. Het.: 27.
1896 *Fleta moorei* – JORDAN, Novit. Zool., 3:35.

Diagnosis: Head, thorax and legs dark chocolate-brown; antennae blackish; palpi, sides of frons, collar mixed with yellowish-white hairs; pectus clothed with yellow hair; legs mixed with whitish scales; abdomen orange yellow with a blackish-brown dorsal stripe and terminal tuft. Fore wings dark chocolate-brown, with an oblique yellow band across the disc, running from below middle of costa to near tornus. Hind wings orange-red with a brownish tinge; the extreme base and a terminal band about 4 mm wide in costal area and $2^1/_2$ – 3 mm towards anal angle, dark brown. Length of fore wing 23 mm.

Distribution: West Java.

### Genus Longicella JORDAN

(fig. 9)

1896 *Longicella* JORDAN, Novit. Zool., 3:51. – Species typica: *Eusemia mollis* WALKER (by original designation).

Diagnosis: Proboscis fully developed; antennae slightly dilated towards the extremity; palpi upturned; 1st and 2nd joints fringed beneath with moderately long hair; 3rd long, naked, porrect; frons with a rounded prominence; tibiae nearly smoothly scaled; abdomen of male with lateral tufts of long hair from the second segment. Venation: in fore wings, vein 2 from three quarters of cell; 3 and 4 well separated; 3, 4 and 5 about equidistant; discocellulars faintly incurved; 6 from lower angle of cell; areole narrow; 7 and the long stalk of 8 + 9 from extremity of areole; 10 from upper margin of areole, near extremity. In hind wings, vein 2 from about middle of cell; 3 and 4 widely apart, the former branching off before the lower angle of cell; cell very long, reaching beyond middle of wing; discocellulars faintly incurved; 5 obsolescent, from middle of discocellulars; 6 and 7 from a point. Male genitalia: Uncus as long as the valva, slender, regularly arched, ending in a short hook. Vincula narrow, bearing each a proximal bunch of long, stiff hair. Valva banana-shaped, but rather broad at base, then narrowing, mainly because of a notable costal incurving; costa faintly outcurved in second third, then slightly in-

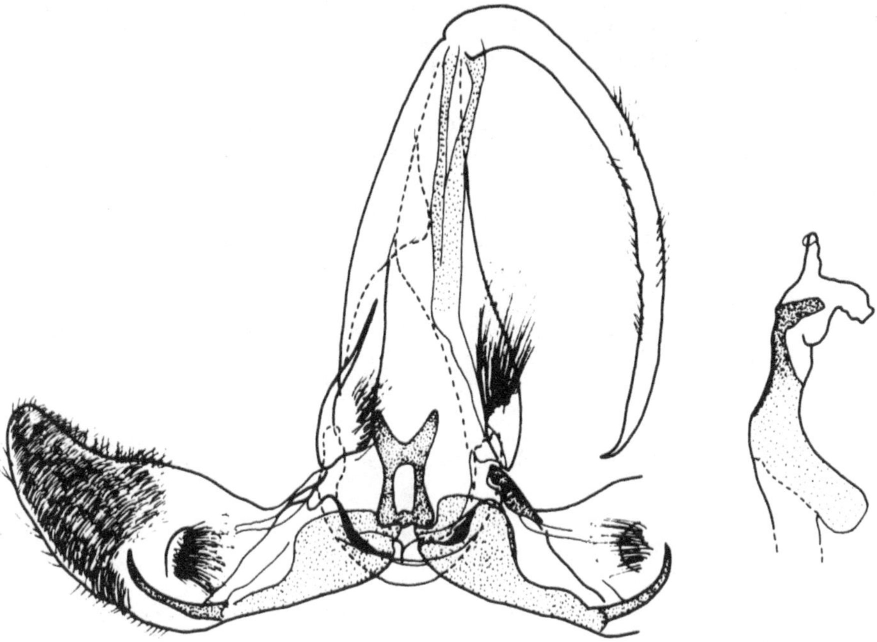

Fig. 9. *Longicella mollis* (WALKER). Type. B. M. Noct. 8507

curved; apex rounded, forming an angle of about 60°; sacculus straight for the basal third, then regularly arched to apex; fold of sacculus very broad, lobe-shaped basally, then quickly narrowing, reaching to one-third of valva and ending in a small, narrow, arched harpe; across the basal third of valvula a transverse ridge bearing a fringe of stiff hair. Aedeagus not longer than the fold of the sacculus, broader basally, nearly straight; fultura inferior produced distally to form a slender, digit-shaped process. Saccus very short, rounded.

Distribution: Malaya, Indonesia.

Key to the species

1 Wings bluish, spotted with black . . . . . . . . . . . . . . . . . . . . . . . . *L. mollis*, p. 45
1′ Wings black, spotted with white . . . . . . . . . . . . . . . . . . . . . . . *L. luctifera*, p. 46

## Longicella mollis (WALKER)

1856 *Eusemia mollis* WALKER, List Lepidopt. Het. Brit. Mus., 7:1774 (♂, ♀). – Terra typica: Malaya (Types, ♂, ♀, in British Museum).
1896 *Longicella mollis* – JORDAN, Novit. Zool., 3:51.
1912 *Longicella mollis* aberr. *catamelas* JORDAN in SEITZ, Großschmetterl. Erde, 11:20 (♂). – Terra typica: Not designated (Type, ♂, in British Museum).

D i a g n o s i s : Body black; 1st and 2nd palpal segments yellowish white; head spotted with yellowish white; collar and tegulae edged with yellowish white hair; pectus with white pilosity; coxae and streaks on legs orange; basal urites edged with white, the rest with orange; underside of abdomen orange. Wings greyish blue with a silvery sheen, sometimes more greenish or more purplish; some scattered black scales; markings black. Fore wings with an irregular subbasal band; a large cell-spot, more or less connected with a roundish spot below cell, both or only the cell-spot with a silvery centre; on discocellulars and beyond a more or less fabiform spot, also centred with silvery; a terminal band of sagittiform spots, connected with a subapical bar; spot in interspace I the largest; silvery spots on termen; cilia concolorous, but white in apical and tornal areas. Hind wings with a basal subcostal spot, a large, nearly square cell-spot, and a very irregular terminal band of lunules, very broad costally; spot in interval I extending to nearly midwing; a series of terminal blue spots; cilia concolorous, tipped with white. Length of fore wing 21 – 25 mm.

D i s t r i b u t i o n : Malaya, Sumatra, Borneo, Poelo Laoet, Nias.

### Longicella mollis mollis (WALKER)

1912 *Longicella mollis mollis* – JORDAN in SEITZ, Großschmetterl. Erde, **11**:20; pl. 4b.

D i s t r i b u t i o n : Malaya, Sumatra, Borneo, Poelo Laoet.

### Longicella mollis decipiens (BUTLER)

1884 *Ophthalmis decipiens* BUTLER, Ann. Mag. Nat. Hist., London, (5) **14**:34 (♂). – T e r r a t y p i c a : Nias Island (Type, ♂, in British Museum).

D i a g n o s i s : Abdomen with the orange colour less extensive; black markings generally more extensive, some veins streaked with black; basal black in fore wings confluent with the medial spots, leaving in the antemedial area only a couple of small blue spots.

D i s t r i b u t i o n : Nias.

### Longicella luctifera (BOISDUVAL)

1836 *Agarista luctifera* BOISDUVAL, Spéc. Gén. Lépidopt., **1**: pl. 14, fig. 4. – T e r r a t y p i c a : Java (Type in British Museum).
1864 *Eusemia luctifera* – WALKER, List Lepidopt. Het. Brit. Mus., **31**:53.
1892 *Phalaenoides luctiferus* – KIRBY, Syn. Cat. Lepidopt. Het.:23.
1896 *Longicella luctifera* – JORDAN, Novit. Zool., **3**:51.

D i a g n o s i s : Black; palpi white laterally at base; head spotted with white; tegulae with white spots; coxae and legs below orange; abdomen with a blue tinge, and the anal tuft orange in male. Fore wings with a subbasal white spot in cell and another below cell; end of cell with a square white spot; a discal oblique white band of three or four spots in interspaces IV–VII, a small spot in interspace II, and a large spot in interspace I, placed more proximally; tornus with a silvery blue spot; other silvery blue marks at base, in cell and on the discocellulars; cilia concolorous, becoming white in apical and tornal areas. Hind wings with a white spot in middle of cell, and another between media and anal margin; a large postdiscal spot in interspace IV, and another in interspace II; cilia as in fore wings. Length of fore wing 25 – 28 mm.

Distribution: Java.

Early stages: Larva velvety black; head and legs orange; each segment bearing a white transverse band. On *Coleus bicolor* (?).

## Genus **Mimeusemia** BUTLER

(fig. 10)

1875 *Mimeusemia* BUTLER, Ann. Mag. Nat. Hist., London, (4) **15**:397. – Species typica: *Mimeusemia persimilis* BUTLER (by original designation).

Diagnosis: Proboscis fully developed; antennae dilated towards the extremity, in female sometimes nearly setiform; palpi obliquely upturned; 2nd joint with long pilosity in front, typically longer at the tip than at the base, so that the joint in lateral view seems to be dilated distally; 3rd joint long, naked, porrect; frons with a truncate conical prominence, with a raised rim at the tip; tibiae nearly smoothly scaled; abdomen with dorsal crests on the basal urites; male with lateral tufts of long

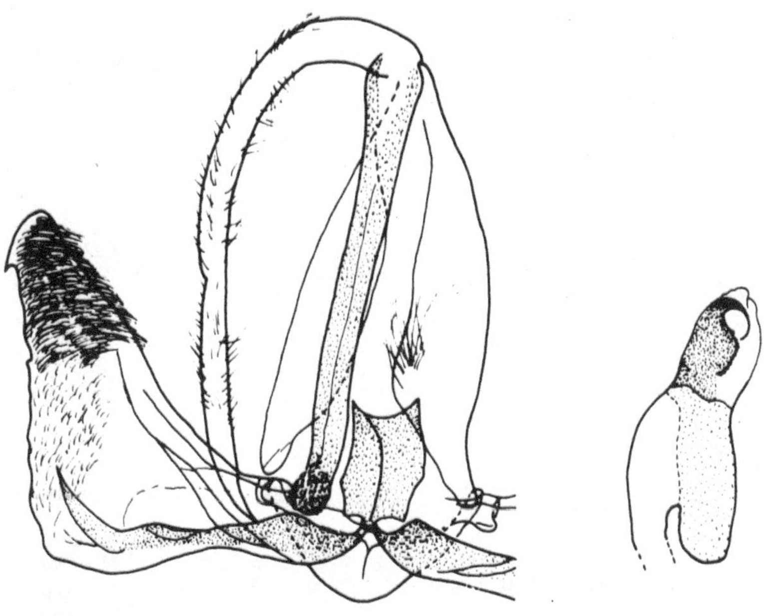

Fig. 10. *Mimeusemia persimilis* BUTLER. B. M. Noct. 8523 (left valva damaged)

hair from the base. Nervation: in fore wings, vein 2 from about five sixths of cell; 3, 4 from a point at the angle of cell; 5 slightly above angle; discocellulars regularly incurved; 6 from the upper cell-angle; areole small and narrow; 7 and the long stalk of 8 + 9 from its extremity; 10 from its upper margin, near to extremity. Hind wings with vein 2 from about five sevenths of cell; 3, 4 and 6, 7 from the cell-angles; discocellulars oblique, very faintly angled inwards at the base of vein 5 which is obsolescent.

Male genitalia: Uncus very long, slightly longer than valva, slender, arched, with a small terminal hook and the pilosity becoming longer towards the extremity. Vinculum moderately broad; proximal half fringed with moderately long, stiff hair. Valva narrow at the base, broadening and leaf-shaped, with the apex produced and nearly acute; harpe placed on a subbasal fold, slender, curved, reaching about halfway the valvula. Aedeagus short, nearly straight, somewhat expanded proximally; fultura inferior little differentiated. Saccus very short, rounded.

Distribution: East and Southeast Asia to India, Indonesia, and Australia.

## Key to the species

1 Hind wings with an orange or creamy-white postmedial patch . . . . . . 2
1′ Hind wings unspotted or with two yellowish spots . . . . . . 17
2 Hind wings with an orange postmedial patch . . . . . . 3
2′ Hind wings with a creamy-white postmedial patch . . . . . . 7
3 Fore wings with the markings orange . . . . . . *M. puciolia*, p. 49
3′ Fore wings with the markings creamy-white . . . . . . 4
4 Fore wings with an oblique band beyond the cell . . . . . . 5
4′ Fore wings without postmedial band, only with two spots . . . . . . 6
5 Abdomen without black bands . . . . . . *M. lombokensis*, p. 49
5′ Abdomen with black bands . . . . . . *M. vittata*, p. 50
6 Hind wings without pale yellowish; black discocellular spot separated from terminal band . . . . . . *M. persimilis*, p. 51
6′ The black discocellular spot connected with terminal band, separating a pale yellowish patch . . . . . . *M. vilemani*, p. 51
7 Abdomen orange, with black dorsal patch at base . . . . . . 8
7′ Abdomen otherwise coloured . . . . . . 10
8 Tegulae spotted with yellow . . . . . . 9
8′ Tegulae without yellow patches . . . . . . *M. postica*, p. 52
9 Hind wings with cilia tipped with white . . . . . . *M. albicilia*, p. 52
9′ Hind wings with cilia tipped with white at the apex only . . . . . . *M. ceylonica*, p. 53
10 Abdomen banded with black and orange . . . . . . 11
10′ Abdomen blackish with the anal tuft orange . . . . . . 16
11 Hind wings with a white basal patch . . . . . . 12
11′ Hind wings without white basal patch . . . . . . 14
12 Hind wings with the white patch extending nearly to the anal angle . . *M. perakana*, p. 53
12′ Hind wings with the white patch extending to middle of wing . . . . . . 13
13 Fore wings with several white spots . . . . . . *M. econia*, p. 53
13′ Fore wings with only two white spots . . . . . . *M. limbata*, p. 54
14 Abdomen with the two basal segments banded with orange . . . . . . *M. peshwa*, p. 55
14′ Abdomen with the two basal segments black . . . . . . 15
15 Abdomen with black bands on medial segments; hind wings with an orange patch usually present on the base of the anal area . . . . . . *M. basalis*, p. 55
15′ Abdomen without black bands on medial segments; hind wings with a creamy patch below the cell . . . . . . *M. davidsoni*, p. 56
16 Fore wings blackish brown . . . . . . *M. accurata*, p. 56
16′ Fore wings chocolate-red partly suffused with ochreous . . . . . . *M. centralis*, p. 57

17 Hind wings unspotted . . . . . . . . . . . . . . . . . . . . . . . . . . . . . . . . . . . . 18
17′ Hind wings with pale yellow medial and postmedial spots . . . . . . . . . . *M. semyron*, p. 57
18 Ground colour chocolate-brown . . . . . . . . . . . . . . . . . . . . . . . . . . *M. simplex*, p. 58
18′ Ground colour black . . . . . . . . . . . . . . . . . . . . . . . . . . . *M. nigrescens*, p. 60

## Mimeusemia puciolia (DRUCE)

1895 *Agarista puciolia* DRUCE, Ann. Mag. Nat. Hist., London, (6) **15**:42 (♂). – Terra typica: Minahasa, Celebes (Type, ♂, in British Museum).
1897 *Mimeusemia fruhstorferi* ROTHSCHILD, Novit. Zool., **4**:183 (♂). – Terra typica: Celebes (Type, ♂, in British Museum).
1901 *Mimeusemia puciolia* – HAMPSON, Cat. Lepidopt. Phal. Brit. Mus., **3**:608; pl. 54, Fig. 3.

Diagnosis: Head and thorax black; antennae with white scales and yellowish white base and tip; palpi yellowish white at base and white in front; sides of frons and streaks on collar, tegulae and middle of thorax yellowish white; throat white; pectus and stripes on legs orange; abdomen orange, with dorsal and lateral black bands extending on last two urites to the underside. Fore wing blackish brown; a whitish basal spot; remaining markings orange, viz.: an oblique antemedial patch from the subcostal to the submedian fold; sometimes connected on the media with the oblique medial band from the subcostal to the submedian fold; a curved band beyond cell from vein 2 to vein 7; there are also some silvery blue markings, viz.: some scales near base below costa and above vein 1, and in middle of cell; a disco-cellular bar with a short streak above it, and some diffused scales below it, and a spot in interspace I near tornus. Hind wings blackish brown; an oblique orange patch from below costa near base to near vein 1 towards tornus, expanding basad below the costa and on the media; an oblique patch beyond cell from below the costa to vein 3; anal margin, termen and cilia towards tornus orange. Length of fore wing 22 – 24 mm.

Distribution: North Celebes (Minahasa = Terra typica).

## Mimeusemia lombokensis ROTHSCHILD

1897 *Mimeusemia lombokensis* ROTHSCHILD, Novit. Zool., **4**:310; pl. 7, fig. 7 (♂, ♀). – Terra typica: Lombok Island (Types, ♂, ♀, in British Museum).
1901 *Mimeusemia lombocensis* – HAMPSON, Cat. Lepidopt. Phal. Brit. Mus., **3**:609.

Diagnosis: Head and thorax black; palpi streaked laterally with pale yellow; sides of frons yellow; collar and tegulae edged with yellow; pectus and legs orange; tarsi banded with black; abdomen orange, with a blackish patch at base dorsally, including the crests. Fore wings deep chocolate-brown, sprinkled with yellow scales; a pale yellow wedge-shaped spot at base of cell; a band across the extremity of cell, reduced sometimes in female to a cell-spot; an oblique band beyond cell between veins 2 and 7, dilated towards dorsum; some silvery-blue markings, viz.: a spot at base of costa; one in cell-middle; a bar on discocellulars; two medial bars below costa to near dorsum; cilia white apically. Hind wings black; an orange patch from near base to beyond the middle and from subcostalis to the anal margin; cilia white apically. Length of fore wing 23 mm.

Distribution: Lombok and East Java.

## Mimeusemia lombokensis lombokensis (ROTHSCHILD)

1912 *Mimeusemis lombokensis lombokensis* – JORDAN in SEITZ, Großschmetterl. Erde, **11**:17; pl. 3e.

Diagnosis: Female with only one cell-spot; band beyond cell narrow, only about 2 mm wide.

Distribution: Lombok (Sapit, 2000 ft.).

## Mimeusemia lombokensis javana (ROTHSCHILD)

1898 *Mimeusemia javana* ROTHSCHILD, Novit. Zool., **5**:36. – Terra typica: Lawang, East Java (Type, ♀, in British Museum).
1912 *Mimeusemia lombokensis javana* – JORDAN in SEITZ, Großschmetterl. Erde, **11**:17.

Diagnosis: Female with two cell-spots; band beyond cell broader.

Distribution: East Java (Lawang).

## Mimeusemia vittata (BUTLER)

1875 *Eusemia vittata* BUTLER Ann. Mag. Nat. Hist., London, (4) **15**:143 (♀). – Terra typica: Java (Type, ♀, in British Museum).
1892 *Episteme vittata* – KIRBY, Syn. Cat. Lepidopt. Het.:27.
1901 *Mimeusemia vittata* – HAMPSON, Cat. Lepidopt. Phal. Brit. Mus., **3**:609; pl. 54, fig. 7.

Diagnosis: Body blackish brown; antennae ringed with white; fore tarsi with white bands; abdomen blackish brown with orange bands. Fore wings deep red-brown; a small yellow antemedial spot below cell; another in cell; an oblique postmedial band from below costa to above tornus; some silvery-blue markings, viz.: subbasal points below costa and on vein 1; a streak below middle of costa; antemedial points in and below cell and above vein 1; a bar on discocellulars and points below the origin of vein 2; a postmedial series of small dentate spots from below costa to vein 3; and some markings on dorsal area towards tornus. Hind wings orange with the base and the costal area brown, the latter expanding into a spot at the upper cell-angle; a terminal brown band expanding widely towards costa and below vein 2. Length of fore wing 26 mm.

Distribution: Java, North Borneo, Malaya.

## Mimeusemia vittata vittata (BUTLER)

Diagnosis: As above.

Distribution: Java and North Borneo.

### Mimeusemia vittata jordani PENDLEBURY

1934 *Mimeusemia vittata* subsp. *jordani* PENDLEBURY, J. Fed. Malay St. Mus., Kuala Lumpur, **17**:566 (♂, ♀). – Terra typica: Kuala Lumpur, Selangor (Types, ♂, ♀, in Federal Malay States Museum).

Diagnosis: Fore wings with a large square creamy-white spot in cell; hind wings with the dark border narrower, especially at the costal margin, giving a much larger free portion to the costal spot.

Distribution: Malaya (Selangor: Kuala Lumpur).

### Mimeusemia persimilis BUTLER

1875 *Mimeusemia persimilis* BUTLER, Ann. Mag. Nat. Hist., London, (4) **15**:397 (♀). – Terra typica: Hakodate, Japan (Type, ♀, in British Museum).

Diagnosis: Body and wings black; 1st and 2nd joints of palpi with pale yellow spots in front at the tip; frons and vertex with a yellow stripe; collar spotted and tegulae striped with yellow; fore coxae orange in front; femora with orange pilosity; hind tibiae ringed with orange; abdomen banded with orange. Fore wings with a square pale yellow antemedial spot in cell; a short streak on subcosta below middle of costa; a squarish spot from middle of cell to submedian fold; a spot beyond cell between veins 6 and 8, variable in size and shape; a triangular spot between veins 3 and 5; some leaden silvery markings, viz.: spots below base of costa, in and below middle of cell; an oblique line across vein 1; a discocellular line and postmedial spots on vein 1 and 2; cilia white apically. Hind wings orange; costal area black, expanding into a patch at base and a square discoidal patch; a terminal band with sinuous proximal edge connected to a spot beyond the lower cell-angle between veins 2 and 4. Length of fore wing 25 – 29 mm.

Distribution: Ussuri region; Japan; Korea; West China.

### Mimeusemia vilemani HAMPSON

1911 *Mimeusemia vilemani* HAMPSON, Ann. Mag. Nat. Hist., London, (8) **8**:414 (♂, ♀). – Terra typica: Kanshirei, Taiwan (Types, ♂, ♀, in British Museum).

Diagnosis: Body black; 2nd joint of palpi white at base and extremity; vertex, collar and tegulae with white stripes; metathorax with a small white spot; pectus orange; tibiae banded with orange; tarsi ringed with white; abdomen with orange bands. Fore wings black; a rounded yellowish white antemedial spot in cell; a medial elliptical spot from discal fold to interval I; oblique postmedial spots beyond the cell-angles, the upper one rounded, the lower triangular; some silvery blue markings, viz.: sprinkling below costa to end of cell; a short streak above base of vein 1; medial spots in and below cell and above vein 1; a discoidal streak; a slight curved postmedial series of spots and two spots on tornus; cilia white apically and on tornus. Hind wings orange; basal, costal and terminal areas black; black apical area expanding squarely into cell to origin of vein 2; a bar-shaped yellowish-white spot on it between veins 4 and 7; cilia with some white towards apex and above vein 1. Length of fore wing 21 – 27 mm.

Distribution: China (Taiwan).

### Mimeusemia postica (WALKER)

1862 *Aegocera postica* WALKER, J. Linn. Soc. London, 6:86 (♂). – Terra typica: Sarawak (Type, ♂, in Museum Oxford University).
1892 *Mimeusemia tara* SWINHOE, Cat. Het. Mus. Oxon., 1:167; pl. 6, fig. 11. (♂). – Terra typica: Sarawak (Type, ♂, in Museum Oxford University).
1901 *Mimeusemia postica* – HAMPSON, Cat. Lepidopt, Phal. Brit. Mus., 3:612.

Diagnosis: Head and thorax chocolate-red; palpi whitish at tips; middle of frons, vertex, stripe on collar and on tegulae pale yellow; pectus and legs orange; tarsi mostly black; abdomen orange with a triangular black dorsal basal patch; last urite with a black patch in male, and with sublateral and anal black spots in female. Fore wings chocolate-red; an antemedial pale yellow cell-spot; medial spots in and below cell; a reniform patch beyond cell between veins 3 and 7; some silvery-blue markings, viz.: some scales at base; antemedial dots in and below cell; a discoidal bar and a spot below it on dorsum; a postmedial line, irregularly interrupted, angled inwards to below end of cell, then excurved again; some scales on termen; cilia white apically. Hind wings black; basal area from costa before middle to tornus orange, with a curved distal edge; a nearly triangular pale yellow spot beyond cell. Length of fore wing 17 – 19 mm.

Distribution: Borneo (Sarawak), Singapore, North Vietnam, coasts of China (Kwangtung), probably very widely occurring.

### Mimeusemia albicilia HAMPSON

1894 *Mimeusemia albicilia* HAMPSON, Moths Ind., 2:160 (♂, ♀). – Terra typica: Tilin Yaw, Burma (Types, ♂, ♀, in British Museum).

Diagnosis: Head and thorax black; base and extremity of the 2nd palpal joint and markings on head pale yellow; middle of collar and tegulae striped with yellow; metathorax with a yellow spot; pectus and legs orange; fore and mid tibiae spotted with black; tarsi with black bands; abdomen orange with black dorsal tufts; a black bar on last urite in male. Fore wings deep chocolate-red in male, blackish brown in female; a pale yellow basal dot; an antemedial spot from the subcostal vein to below cell; a medial spot on the subcostal vein; a nearly triangular spot from discal fold in cell to submedian fold; a patch beyond cell from vein 10 to just below vein 3; some silvery-blue markings, viz.: subbasal spots; spots in and below cell and on vein 1; a discoidal bar; a postmedial line; a few spots in the tornal area; cilia white apically and on tornus. Hind wings blackish brown; a pale orange patch on anal area, from base to beyond middle, and at base to costa; a pale yellow rounded spot beyond the cell, between veins 3 and 7; cilia tipped with white. Length of fore wing 23 – 26 mm.

Distribution: Burma.

## Mimeusemia ceylonica HAMPSON

1893 *Mimeusemia ceylonica* HAMPSON, Illustr. Het. Brit. Mus., 9:89; pl. 157, fig. 24 (♀). – Terra typica: Trincomali, Ceylon (Type, ♀, in British Museum).

Diagnosis: Head and thorax blackish brown; base and extremity of the 2nd palpal joint, frons, and vertex pale yellow; teguale with a yellow patch; pectus and legs orange; fore and mid tibiae spotted with black; tarsi ringed with black; abdomen orange; a triangular black patch at base and a sublateral series of four spots. Fore wings deep chocolate brown; a triangular yellowish white antemedial patch in and below cell; a medial point or streak on the subcostalis; an oblique patch from discal fold in cell to the submedian fold; an oblique, roughly lunulate patch beyond cell between veins 3 and 8; some silvery-blue markings, viz.: diffused streaks of scales below basal half of costa and above vein 1; a spot in middle of cell; a discoidal bar; a postmedial line, interrupted in places, curved from costa to vein 3, then retracted; termen sprinkled with silvery-blue scales; cilia more or less white apically and on tornus. Hind wings blackish brown; an orange patch on anal area to beyond middle and at base to costa; a squarish yellowish white spot beyond cell, between veins 3 and 7; cilia tipped with white. Length of fore wing 20 mm.

Distribution: Ceylon.

## Mimeusemia perakana ROTHSCHILD

1896 *Mimeusemia perakana* ROTHSCHILD, Novit. Zool., 3:50; pl. 14, fig. 19 (♀). – Terra typica: Sandakan, Borneo (Type, ♀, in British Museum).
1901 *Mimeusemia peracana* HAMPSON, Cat. Lepidopt. Phal. Brit. Mus., 3:612.

Diagnosis: Head and thorax deep chocolate-red; base and extremity of the second palpal joint, a broad stripe from frons over collar and thorax, and stripes on tegulae, white; pectus and legs orange; fore and mid tibiae spotted with black; tarsi ringed with black; abdomen orange with the base blackish, and with segmental black bands. Fore wings deep chocolate-red; an oblique white antemedial patch from near costa to submedian fold; oblique conjoined medial spots from near costa to interval I; a lunulate patch beyond cell between veins 3 and 8; some silvery blue markings, viz.: some scales near base; antemedial spots in and below cell and on vein 1; a discoidal bar; a sinuous postmedial line; termen mixed with silvery blue scales; cilia white apically. Hind wings blackish brown; a yellowish white patch on anal area from base to beyond middle, extending to the subcostal vein near base; cilia white. Length of fore wing 23 mm.

Distribution: Perak, Borneo (Sandakan). Java (?).

## Mimeusemia econia HAMPSON

1900 *Mimeusemia econia* HAMPSON, Mon. Christmas Island 64; pl. 9, fig. 7 (♂, ♀). – Terra typica: Christmas Island (Types, ♂, ♀, in British Museum).

D i a g n o s i s : Head and thorax black; base and extremity of the second palpal joint, sides of frons, and vertex yellowish white; collar, tegulae and thorax striped with yellowish; pectus and legs orange; fore and mid tibiae spotted with black; tarsi ringed with black; abdomen orange with a diffused dorsal black band. Fore wings chocolate-red; terminal area sprinkled with yellowish scales; a yellowish-white point at base; an antemedial triangular spot from the subcostalis to submedian fold; a medial spot on the subcostalis; a spot from above the media to submedian fold; an oblique spot beyond cell from interspace III to vein 8; some silvery blue markings, viz.: streaks below the basal half of costa and of vein 1; spots in and below middle of cell; a discoidal bar; a postmedial line, curved from costa to vein 3, then inwards; cilia yellowish at apex and tornus. Hind wings blackish-brown; a yellowish patch on basal half of the anal area, to the subcostalis; an elliptical patch beyond cell between veins 3 and 7; cilia whitish apically and below vein 2. Length of fore wing 18 – 20 mm.

D i s t r i b u t i o n : Christmas Island; West Celebes; West Bali.

### M i m e u s e m i a e c o n i a e c o n i a (HAMPSON)

D i a g n o s i s : As above.

D i s t r i b u t i o n : Christmas Island.

### M i m e u s e m i a e c o n i a h e d y a JORDAN

1939 *Mimeusemia econia hedya* JORDAN, Novit. Zool., **12**:248 (♂, ♀). – T e r r a t y p i c a : Western Celebes (Types, ♂, ♀, in British Museum).

D i a g n o s i s : Fore tarsi with the first ring yellow, remaining segments black with a white apical spot; mid and hind tarsi black with a few white scales at the segmental tips; abdomen black above and at sides, not banded with yellow; underside with a broad yellow stripe from base to the yellow apical tuft; fore wings with the spots purer white, as a rule a little larger, especially the median spot, which is irregularly rounded.

R e m a r k : In view of the slight differences in the genitalia, this form might be considered as a distinct species. – Subsp. 1 of HAMPSON (Cat. Lepidopt. Phal. Brit. Mus., **3**:613): 'Hind wing with the yellowish patch reduced and not extending to inner margin' = var. *regina* STRAND (Lepidopt. Cat., **5**:26, Queensland; coll. LOWER), probably is not conspecific. Should it prove a valid species, the name *regina* STRAND would be available to it.

D i s t r i b u t i o n : West Celebes.

### **Mimeusemia limbata** JORDAN

1939 *Mimeusemia limbata* JORDAN, Novit. Zool., **41**:248–249 (♂, ♀). – T e r r a t y p i c a : West Celebes (Types, ♂, ♀, in British Museum).

D i a g n o s i s : Head and thorax black; base and extremity of the 2nd palpal joint yellowish white; collar, tegulae and thorax striped with yellowish; pectus and legs orange; hind tibiae and all tarsi black; tarsal joints white at the extremities, except the first joint of the fore tarsi which is orange; hind tibiae with a submedian spot and the tip yellow; abdomen black with the extremity orange; in male, the basal quarter orange on underside; in female, a white lateral basal spot. Wings black. Fore wings with a creamy-white subcostal dot near the base; an antemedial oblique spot touching the base of the second media; a small spot at the upper cell-angle; a discal spot somewhat larger than the antemedial one, not reaching costa nor the dorsal area; some silvery blue markings, viz.: a short streak below cell; a larger one along dorsum; a narrow streak above vein 2; a broader one below vein 2; cilia white apically and above tornus. Hind wings with a creamy-white rounded spot on costa beyond the cell-tip, and a larger patch in front of the anal margin, entering cell, extreme base and cilia along the anal margin black; cilia white at apex and before tornus, elsewhere mixed with black and white. Length of fore wings 11 – 11,5 mm.

D i s t r i b u t i o n : West Celebes.

### Mimeusemia peshwa (MOORE)

1859 *Eusemia peshwa* MOORE, Lepidopt. East Ind. Co., **2**:289; pl. 7, fig. 2 (♀). – T e r r a t y p i c a : Sikkim (Type, ♀, in British Museum).
1894 *Mimeusemia peshwa* – HAMPSON, Moths India, **2**:159.
1901 *Mimeusemia peshwa* – HAMPSON, Cat. Lepidopt. Phal. Brit. Mus., **3**:613.

D i a g n o s i s : Head, thorax and wings blackish-brown; palpi with the 2nd joint yellow at base and extremity; frons, vertex, middle of collar and tegulae striped with yellow; pectus and legs orange; fore and mid tibiae spotted with brown; tarsi banded with brown; abdomen orange with brown dorsal crests and segmental dorsal black bands. Fore wings with a yellow basal dot; a small square antemedial cell-spot; sometimes a medial dot on the subcostal vein; an oblique patch from discal fold in cell to submedian fold; a more or less lunulate patch beyond cell between veins 3 and 8; some silvery-blue markings, viz.: diffused streaks below basal half of costa and in interval I; a spot in middle of cell; a discoidal bar; a postmedial series of spots, sometimes fused, curved from costa to vein 3, then strongly retracted at vein 2; some spots near tornus; cilia white apically. Hind wings with an orange patch in anal area, extending at the base to middle of cell, and on the anal margin to beyond middle; a square yellow patch beyond cell between veins 3 or 4 and 7; cilia white apically. Length of fore wing 24 – 29 mm.

D i s t r i b u t i o n : West Pakistan; North India; Sikkim.

### Mimeusemia basalis (WALKER)

1854 *Eusemia basalis* WALKER, List Lepidopt. Het. Brit. Mus., **1**:53 (♂, ♀). – T e r r a t y p i c a : Northeastern Bengal (Types, ♂, ♀, in British Museum).
1901 *Mimeusemia basalis* – HAMPSON, Cat. Lepidopt. Phal. Brit. Mus., **3**:613 (♂, ♀).

1908 *Mimeusemia hainana* JORDAN Novit. Zool., **15**:254; pl. 9, fig. 10 (♂). – Terra typica: Hainan (Type, ♂, in British Museum).
1912 *Mimeusemia basalis* aberr. *siccima* STRAND, Lepidopt. Cat., **5**:26. – Terra typica: Sikkim.
1916 *Mimeusemia basalis* aberr. *feminalis* STRAND, Arch. Naturgesch., **A2**:3861. – Terra typica: Sikkim.

Diagnosis: Body and wings blackish-brown; abdomen with dorsal bands of orange on urites IV–VII; underside orange except at the extremity. Fore wings with a roughly triangular pale yellow antemedial cell-spot extending above and below, but varying in size; a postmedial band from below costa to submedian fold, angled inwards below vein 4, and with waved edges; some silver-blue markings, viz.: a spot below base of costa; a streak on base of vein 1; a bar in middle of cell; spots on origin of veins 1 and 2; a discoidal bar; a curved postmedial series of elongate spots on veins; a terminal spot above tornus. Hind wings with an orange patch on basal half of dorsal area, very variable in size, sometimes absent, sometimes connected to a large elliptical pale yellow spot beyond cell between veins 2 and 6. Length of fore wing 25 – 29 mm.

Early stages: Larva greyish yellow, tinged with greenish on the first three segments; head orange, brownish at sides, with black points and sutures; warts, streaks on the folds, and hair brown. On *Dillenia*.

Distribution: Bengal; Sikkim; Hainan. The Hainan specimens seem not to differ from the continental ones, which presupposes a distribution across Burma, Indo-China and South China.

## Mimeusemia davidsoni SWINHOE

1899 *Mimeusemia davidsoni* SWINHOE, Ann. Mag. Nat. Hist., London (7) **3**:112 (♀). – Terra typica: Karwar, India (Type, ♀, in Museum Oxford University).

Diagnosis: Body and wings black; abdomen with dorsal orange bands on urites IV–VII; underside orange except at the extremity. Fore wings with a roughly triangular pale yellow antemedial spot from above and below cell; an oblique postmedial band from below costa to submedian fold, its proximal margin excised beyond the lower cell-angle, its distal margin excurved and crenulate below vein 5; some silvery-blue spots, viz.: some basal spots; a spot below middle of costa; a curved series in cell, on the median vein, below cell, above and on vein 1; a discoidal bar; a postmedial spot on vein 1; a subterminal series of streaks on veins, ending in a transverse mark near tornus; some terminal sprinkling. Hind wings with a pale yellow spot below cell-end, conjoined to a spot beyond cell. Length of fore wing 25 mm.

Distribution: South India: Kanara.

## Mimeusemia accurata (SWINHOE)

1890 *Eusemia accurata* SWINHOE, Proc. Zool. Soc. London, **1889**:401 (♀). – Terra typica: Moulmein, Burma, (Type, ♀, in British Museum).
1892 *Episteme accurata* – KIRBY, Syn. Cat. Lepidopt. Het.:30.
1901 *Mimeusemia accurata* – HAMPSON, Cat. Lepidopt. Phal. Brit. Mus., **3**:614; pl. 54, fig. 23.

D i a g n o s i s: Body and wings blackish-brown; palpi marked with white; frons and vertex with a yellowish-white stripe; collar with a yellowish median spot; tegulae with a yellowish stripe; pectus, coxae, femora and tibiae orange; abdomen with the anal tufts and ventral bands orange. Fore wings with a yellowish white point at cell-base; an antemedial bar across the cell; a medial band from below costa to above vein 1, constricted in cell and expanding greatly on the outer side below cell; a broad band beyond cell between veins 3 and 8, its distal edge excised at middle; some silvery-blue spots, viz.: a spot below costa near base; antemedial spots in and below cell and above vein 1; a discoidal bar; a postmedial line, strongly incurved below vein 4 and oblique from vein 2 to dorsum; cilia white apically. Hind wings with a rounded yellowish white spot beyond cell between veins 3 and 7; cilia white at tips. Length of fore wing 29 mm.

D i s t r i b u t i o n: Burma (Moulmein = terra typica).

## Mimeusemia centralis (ROTHSCHILD)

1896 *Phalaenoides centralis* ROTHSCHILD, Novit. Zool., **3**:326 (♀). – T e r r a t y p i c a: Rockhampton, Queensland (Type, ♀, in British Museum).
1901 *Mimeusemia centralis* – HAMPSON, Cat. Lepidopt. Phal. Brit. Mus., **3**:615; pl. 51, fig. 11.

D i a g n o s i s: Head yellow; palpi and area behind the antennae with some red-brown; thorax red-brown; collar, tegulae and middle of thorax with yellow stripes; pectus and legs orange; tibiae and tarsi banded with brown above; abdomen blackish-brown; two last urites and underside orange. Fore wings chocolate-red, tinged with ochreous, except on the basal dorsal area, in cell and on veins; an oblique wedge-shaped yellowish antemedial band from the subcostal vein to vein 1; a medial band across cell to just below it; an oblique postmedial band with rounded extremities, from below vein 3 to vein 7; some silvery markings, viz.: scales below base of costa; on vein 1; below middle of cell; on and below the discocellulars; traces of a postmedial line incurved below vein 3. Hind wings chocolate-brown, with a rounded yellowish spot beyond cell; some yellowish scales on median part of the terminal area; cilia yellowish. Length of fore wing 26 mm.

D i s t r i b u t i o n: North Queensland.

E a r l y s t a g e s: Larva pale yellow, ringed with black; head, prolegs and a series of spots above the legs, orange; first thoracal segment black dorsally; a transversal spot on the eleventh segment black.

## Mimeusemia semyron (HERRICH–SCHÄFFER)

1853 *Agarista semyron* HERRICH–SCHÄFFER, Außereurop. Schmetterl.: fig. 24. – T e r r a t y p i c a: Philippine Islands (Type in Berlin Museum).
1866 *Eusemia chrysospila* WALKER, List. Lepidopt. Het. Brit. Mus., **35**:1860 (♂). – T e r r a t y p i c a: Philippine Islands (Type, ♂, in British Museum).
1874 *Eusemia semyron* – BOISDUVAL, Rev. Mag. Zool., (3) **2**:96.
1892 *Phalaenoides semyron* – KIRBY, Syn. Cat. Lepidopt. Het.:24.
1901 *Tuerta semyron* – HAMPSON, Cat. Lepidopt. Phal. Brit. Mus., **3**:627 (♂, ♀).
1912 *Mimeusemia semyron* – JORDAN in SEITZ, Großschmetterl. Erde, **11**:17; pl. 3g.

D i a g n o s i s: Head, thorax and wings blackish-brown; palpi pale yellow at base and in front; sides of frons and a spot on back of head yellow; postocular line yellow; middle of collar, shoulders and tegulae striped with yellow; metathorax with a yellow spot; pectus and legs orange; fore and mid tibiae with black spots; tarsi ringed with black; abdomen orange, banded with black. Fore wings with a pale yellowish point at base; an antemedial and a medial square spots in cell; a square spot below cell-extremity; spots beyond cell between veins 3 and 5, and 5 and 8; some silvery blue markings, viz.: markings at base; antemedial spots in and below cell and above vein 1; a discoidal bar; spots on dorsal margin beyond middle and near tornus; cilia white apically and at tornus. Hind wings with a pale yellowish cell-spot, connected with a square spot below cell; a rounded spot beyond the cell between veins 3 and 7; cilia white at apex and below vein 1. Length of fore wing 20 – 26 mm.

D i s t r i b u t i o n: Philippine Islands: Luzon; Sumatra (very doubtful).

## **Mimeusemia simplex** (LUCAS)

1891 *Agarista simplex* LUCAS, Proc. Linn. Soc. New South Wales, Sydney, (2) 8:140 (♂). – T e r r a t y p i c a: North Queensland (Type, ♂, in South Australian Museum Adelaide).
1912 *Mimeusemia simplex* – JORDAN in SEITZ, Großschmetterl. Erde, **11**:18.

D i a g n o s i s: Body blackish-brown; marks on palpi and head yellowish white, variable; pectus in front and coxae orange; anal tuft of abdomen orange. Fore wings chocolate-brown; a small yellowish spot in cell, sometimes absent; a pale yellow discoidal band beyond cell, from vein 1 to 7; some silvery-blue markings, viz.: some scales below basal half of costa, and others on the veins; a spot in middle of cell; a discoidal bar; extremities of veins with short blue streaks. Hind wings blackish tinged with chocolate; cilia whitish yellow, wholly or partially, expanding sometimes on terminal area, especially before tornus. Length of fore wings 29 mm in an average.

D i s t r i b u t i o n: Moluccas and New Guinea; North Queensland (terra typica).

### M i m e u s e m i a s i m p l e x s i m p l e x (LUCAS)

1912 *Mimeusemia simplex simplex* – JORDAN in SEITZ, Großschmetterl. Erde, **11**:18.

D i a g n o s i s: As above.

D i s t r i b u t i o n: North Queensland.

### M i m e u s e m i a s i m p l e x e u m e l a s JORDAN

1912 *Mimeusemia simplex eumelas* JORDAN in SEITZ, Großschmetterl. Erde, **11**:18; pl. 3f (♂, ♀). – T e r r a t y p i c a: Gilolo Island (Type, ♂, in British Museum).
1939 *Mimeusemia eumelas* – JORDAN, Novit. Zool., **41**:249; fig. 224 (♂).

D i a g n o s i s: Differs as follows from the nymotypical race: creamy band on the fore wings narrow, not quite reaching vein 1, and frequently followed by an isolated spot below vein 1; cilia of hind wings entirely white. Halmaheira. There are some differences in the male genitalia, which induced Jordan to grant this form the species status.

D i s t r i b u t i o n: Halmaheira.

## Mimeusemia simplex obiensis (ROTHSCHILD)

1898 *Mimeusemia obiensis* ROTHSCHILD, Novit. Zool., **5**:418 (♀). – T e r r a t y p i c a: Obi Island (Type, ♀, in British Museum).
1912 *Mimeusemia simplex obiensis* – JORDAN in SEITZ, Großschmetterl. Erde, **11**:18.

D i a g n o s i s: Palpi with the 2nd joint at tip and the 3rd at base yellowish white; pale streaks on head and thorax broad. Fore wings with the antemedial and medial creamy bars across the cell, or the medial bar only, the antemedial bar being in that case short; postmedial band wide, not quite reaching the submedia, and twice excised proximally. Hind wings with only the cilia white.

D i s t r i b u t i o n: Obi Island.

## Mimeusemia simplex buruensis JORDAN

1912 *Mimeusemia simplex buruensis* JORDAN in SEITZ, Großschmetterl. Erde, **11**:18 (♂). – T e r r a t y p i c a: Buru Island (Type, ♂, in British Museum).
1901 *Mimeusemia proerorsia* subsp. *obiensis*/HAMPSON, Cat. Lep. Phal. Brit. Mus., **3**:616.

D i a g n o s i s: Differing from *M. s. obiensis* in the proximal cell-spot which is minute or absent.

D i s t r i b u t i o n: Boeroe Island.

## Mimeusemia simplex proerosia (DRUCE)

1893 *Agarista proerosia* DRUCE, Ann. Mag. Nat. Hist., London, (6) **15**:41 (♂). – T e r r a t y p i c a: Key Island (Type, ♂, in British Museum).
1901 *Mimeusemia proerosia* – HAMPSON, Cat. Lep. Phal., **3**:615; pl. 54, fig. 1.
1912 *Mimeusemia simplex proerosia* – JORDAN in SEITZ, Großschmetterl. Erde, **11**:18; pl. 3g.

D i a g n o s i s: Fore wings without yellowish cell-spots; postmedian band nearer to cell; blue discal line broken into spots and placed more proximally. Hind wings with the termen white in the anal region. Key Islands.

D i s t r i b u t i o n: Key Islands.

### Mimeusemia simplex vitticollis ROTHSCHILD

1896 *Mimeusemia vitticollis* ROTHSCHILD, Novit. Zool., **3**:94 (♂). – Terra typica: Fergusson Island (Type, ♂, in British Museum).
1912 *Mimeusemia simplex vitticollis* – JORDAN in SEITZ, Großschmetterl. Erde **11**:18.

Diagnosis: Yellowish stripes on collar, tegulae and thorax distinct; fore wings with much less blue scales, especially the postmedial line mostly absent; hind wings with a few white terminal scales before the cilia.

Distribution: Fergusson Island (Terra typica); Irian; Papua.

### Mimeusemia simplex limata JORDAN

1926 *Mimeusemia simplex limata* JORDAN, Novit. Zool., **33**:359 (♂, ♀). – Terra typica: New Britain (Type, ♂, in British Museum).

Diagnosis: Fore wings with conspicuous continuous blue discal line, and another stripe along the dorsum, with a third on vein 2; the creamy white band not reaching vein 2, narrower costally; female with a subbasal creamy white dot in cell and two dots beyond middle of cell; hind wings with white of the cilia penetrating the wing before tornus.

Distribution: New Britain (Terra typica), New Ireland.

### Mimeusemia nigrescens JOICEY & TALBOT

1916 *Mimeusemia nigrescens* JOICEY & TALBOT, Trans. Entomol. Soc. London, **1916**:81 (♀). – Terra typica: Schouten Island (Type, ♀, in British Museum).

Diagnosis: Body black; 1st palpal joint fringed with pale yellow; 2nd pale yellow on inner side; a pale yellow line above the eyes; pectus and femora orange; collar with a pale yellow line; tegulae edged with pale yellow on inner side; anal tuft orange. Fore wings black; a nearly cream-coloured discal band from the subcostal vein to near tornus, constricted slightly towards costa and rounded towards dorsum; at its lower distal edge, a curved pale yellow mark; some metallic blue markings, viz.: a cell-spot; a stripe on the discocellulars; an incurved series of five dots just beyond the band; some scales along the costal margin of cell and along the lower submedia. Hind wings black, with pale yellow marginal patches; cilia yellow, blackish proximally. Length of fore wing 27 mm.

Distribution: Schouten Islands.

## Genus Immetalia JORDAN

(fig. 11)

1896 *Immetalia* JORDAN, Novit. Zool., **3**:31. – Species typica: *Eusemia saturata* WALKER (by original designation).

D i a g n o s i s: Proboscis fully developed; 1st and 2nd joints of palpi roughly scaled, 3rd elongate, porrect, prismatic in female; antennae distinctly dilated towards extremity; frons convex, without prominence, or a very weak prominence, rarely a raised rim; hind tibiae with some long hair on outer side towards extremity. Venation: Fore wing with vein 2 from near angle of cell; 3, 4 and 5 approximated, equidistant; discocellulars deeply incurved; 6 from the extreme base of the very narrow areole; 7 and 10, 8 + 9 from extremity of areole. Hind wings with vein 2 from five sixths of cell; 3, 4 and 6, 7 from angles of cell; discocellulars incurved; 5 obsolescent, from middle of discocellulars.

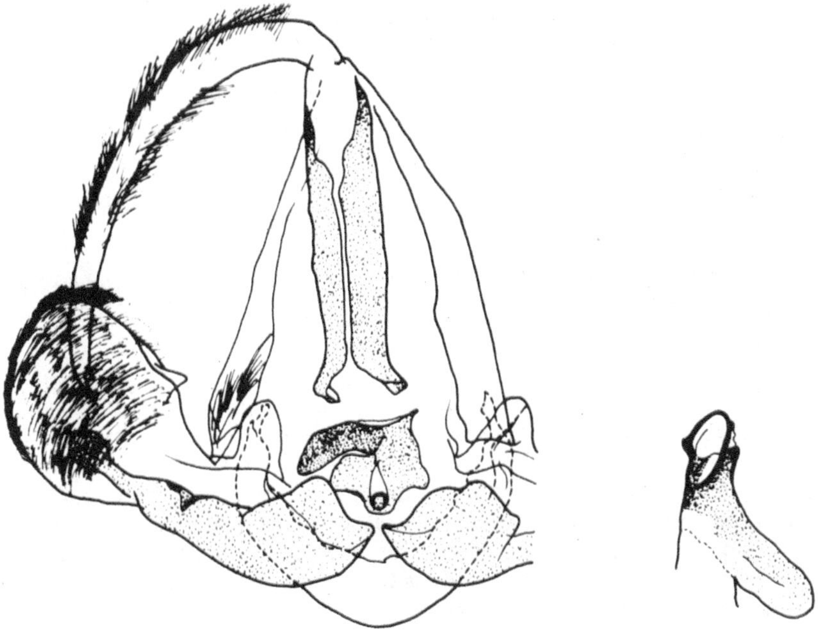

Fig. 11. *Immetalia saturata* (WALKER). B. M. Noct. 8506

Male genitalia: Uncus slightly longer than sacculus, slender, nearly regularly curved, with a tapering extremity. Tubus analis about two-thirds of uncus in length, narrow, cut square at the tip. Vincula narrowing distally, bearing proximally a strong bunch of very long and stiff hair. Valva strongly narrowed at one quarter of sacculus, then broadly leaf-shaped; costa arched, bearing at middle a small process; apex obtuse; sacculus outcurved at base, then faintly incurved, finally broadly arched to apex; fold broad proximally, then narrowing, reaching to two-fifths of sacculus and ending in a short, broad, curved and slightly hooked harpe. Aedeagus somewhat shorter than fold of sacculus, rather strong, nearly straight, bearing terminally a small, slender, nearly filiform process; fultura inferior narrow, but broadening distally to end in two divergent, triangular lobes. Saccus nearly triangular.

Distribution: Indonesia, New Guinea, Bismarck Archipelago, Solomon Islands.

Key to the species

1 Hind wings unspotted . . . . . 2
1' Hind wings with orange or white markings . . . . . 5
2 Fore wings unspotted . . . . . *I. cyanea*, p. 66
2' Fore wings with an orange or reddish band . . . . . 3
3 Band of fore wings short: distance between its extremities and costa and especially tornus, conspicuous . . . . . *I. saturata*, p. 63
3' Band of fore wings longer, especially towards the costa . . . . . 4
4 Band of fore wings broad, orange . . . . . *I. celebensis*, p. 66
4' Band of fore wings narrower, orange red, nearly reaching tornus . . . . . *I. eichhorni*, p. 67
5 Markings of head, palpi and wings orange; band of fore wings rather broad, excurved; hind wings with the black basal portion rather broad . . . . . *I. bernsteini*, p. 62
5' Markings of head and palpi white; markings of wings orange or white . . . subspecies of *I. saturata*, p. 64–66

## Immetalia bernsteini (VOLLENHOVEN)

1863 *Ophthalmis bernsteini* VOLLENHOVEN, Tijdschr. Entomol.'s Gravenhage, Amsterdam, 6:132; pl. 9, fig. 1. – Terra typica: Moluccas (Type not designated).
1892 *Episteme bernsteini* – KIRBY, Syn. Cat. Lepidopt. Het.:28.
1896 *Immetalia bernsteini* – JORDAN, Novit. Zool., **3**:32.

Diagnosis: Body and wings black; palpi mostly orange in front; frons tinged with blue; vertex and throat mixed with orange; extremity of abdomen orange except at extreme tip. Fore wing with an oblique band of orange, rather irregular, running from costa to vein 2, sometimes to vein 1; silvery blue scales forming streaks below costa to middle of wing, on vein 1, and a bar across extremity of cell; cilia concolorous, but tinged with rufous apically. Hind wings with a broad orange postmedial band, from below costa to anal margin, sometimes narrowing towards the anal angle. Length of fore wing 28 mm.

Distribution: Moluccas.

### Immetalia bernsteini bernsteini (VOLLENHOVEN)

1864 *Eusemia josioides* WALKER, List Lepidopt. Het. Brit. Mus., **31**:54 (♂). – Terra typica: Morotai Island (Type, ♂, in Museum Oxford University).
1874 *Agarista xanthomelas* BOISDUVAL, Rev. Mag. Zool., (3) **2**:101. – Terra typica: Morotai Island (Type in British Museum).
1892 *Episteme josioides* – KIRBY, Syn. Cat. Lepidopt. Het.:28.
1912 *Immetalia bernsteini bernsteini* aberr. *brevimaculata* JORDAN in SEITZ, Groß-schmetterl. Erde, **11**:10 (♂, ♀). – "Terra typica": Morotai Island (Types, ♂, ♀, in British Museum).

Diagnosis: As above.

Distribution: Morotai Island (Terra typica), Halmahera, Gilolo.

### Immetalia bernsteini angustiplaga ROTHSCHILD

1896 *Immetalia bernsteini angustiplaga* ROTHSCHILD, Novit. Zool., **3**:32 (♂, ♀). – Terra typica: Batjan Island (Types, ♂, ♀, in British Museum).

Diagnosis: Second joint of palpi wholly or nearly so orange beneath; cilia of fore wings always or nearly always yellow; hind wings with the orange band narrower than the basal black portion.

Distribution: Batjan Island.

### Immetalia bernsteini obiana JORDAN

1912 *Immetalia bernsteini* subsp. *obiana* JORDAN in SEITZ, Großschmettlerl. Erde, **11**:10 (♀). – Terra typica: Obi Island (Type, ♀, in British Museum).

Diagnosis: Second joint of palpi nearly pure orange; yellow band of hind wings narrower, about half as broad as the basal black area; cilia of both wings yellow apically.

Distribution: Obi Island.

### Immetalia bernsteini buruana JORDAN

1912 *Immetalia bernsteini* subsp. *buruana* JORDAN in SEITZ, Großschmetterl. Erde, **11**:10 (♀). – Terra typica: Buru Island (Type, ♀, in British Museum).

Diagnosis: Second joint of palpi streaked with black below; band of fore wings broader than in *I. b. angustiplaga*; hind wings as in that race, but the black margin narrowing towards costa.

Distribution: Boeroe Island.

### **Immetalia saturata** (WALKER)

1864 *Eusemia saturata* WALKER, List Lepidopt. Het. Brit. Mus., **31**:54 (♂). – Terra typica: Boeroe Island (Type, ♂, in Museum Oxford University).
1892 *Episteme saturata* – KIRBY, Syn. Cat. Lepidopt. Het.:28.
1896 *Immetalia saturata* – JORDAN, Novit. Zool., **3**:31.

Diagnosis: Body and wings brownish black with blue reflections; base of palpi, a line on frons, throat and a ring round the neck white; anal tuft orange. Fore wing with a curved orange or white band across and beyond cell from below costa to above tornus, narrowing towards both extremities and with waved edges; cilia white apically. Hind wings unmarked, sometimes with a white or orange band; cilia white apically. Length of fore wing 24 – 28 mm.

Distribution: Moluccas, eastwards to the Bismarck Archipelago.

## Immetalia saturata saturata (WALKER)

1912 *Immetalia saturata saturata* – JORDAN in SEITZ, Großschmetterl. Erde, **11**:10; pl. 2a.

Diagnosis: Band of fore wing orange; hind wings unmarked.

Distribution: Boeroe.

## Immetalia saturata doleschalli (FELDER)

1874 *Agarista doleschalli* FELDER, Reise Novara: pl. 107, fig. 3. – Terra typica: Amboina (Type in British Museum).
1874 *Eusemia doleschalli* – BOISDUVAL, Rev. Mag. Zool., (3) **2**:102.
1892 *Episteme doleschalli* – KIRBY, Syn. Cat. Lepidopt. Het.:28.
1901 *Immetalia saturata* subsp. *doleschalli* – HAMPSON, Cat. Lepidopt. Phal. Brit. Mus., 3:537.

Diagnosis: Fore wings with the median band white in male, orange in female; hind wings without band.

Distribution: Amboina.

## Immetalia saturata leucomelas JORDAN

1912 *Immetalia saturata leucomelas* JORDAN in SEITZ, Großschmetterl. Erde, **11**:10 (♂, ♀). – Terra typica: Ceram Island (Types, ♂, ♀, in British Museum).

Diagnosis: Fore wing with the median band white in both sexes, reaching in male beyond the submedia.

Distribution: Ceram.

## Immetalia saturata keiana ROTHSCHILD & JORDAN

1903 *Immetalia saturata* subsp. *keiana* ROTHSCHILD & JORDAN, Novit. Zool., **10**:486 (♂). – Terra typica: Groot-Kei Island (Type, ♂, in British Museum).
1920 *Immetalia meeci* subsp. *ceiana* HAMPSON (ex errore pro *keiana*), Cat. Lepidopt. Phal. Brit. Mus., Suppl. **2**:538.

Diagnosis: Both wings in male with an orange band; band of fore wings narrowing at both extremities, lunulate, broadest at middle; band of hind wings narrower than the basal black area, not reaching the subcostal vein.

Distribution: Great Kei Island.

## Immetalia saturata longipalpis (KIRSCH)

1877 *Eusemia longipalpis* KIRSCH, Mitt. Mus. Dresden, **1**:130; pl. 130, fig. 12. – Terra typica: New Guinea (Type in Dresden Museum).
1880 *Agarista bruijni* OBERTHÜR, Ann. Mus. civ. Stor. Nat. Genova, **15**: pl. 4, fig. 6 (♀). – Terra typica: New Guinea (Type, ♀, in British Museum).

1886 *Eusemia bruijni* – PAGENSTECHER, Jahrb. Nassau. Ver. Naturk., Wiesbaden, **39**:118.
1892 *Episteme longipalpis* – KIRBY, Syn. Cat. Lepidopt. Het.:28.
1892 *Phalaenoides bruijni* – KIRBY, Syn. Cat. Lepidopt. Het.:23.
1892 *Eusemia doreyana* SWINHOE, Cat. Het. Mus. Oxon.:164; pl. 5, fig. 4 (♀). – Terra typica: New Guinea (Type, ♀, in Museum Oxford Univ.).
1892 *Agarista cynapes* DRUCE, Ann. Mag. Nat. Hist., London, (6) **14**:22 (♂, ♀). – Terra typica: New Guinea (Types, ♂, ♀, in British Museum).
1892 *Immetalia prochyta* DRUCE, Ann. Mag. Nat. Hist., London, (6) **14**:23. – Terra typica: New Guinea (Type in British Museum).
1896 *Immetalia longipalpis* – JORDAN, Novit. Zool., **3**:32.
1912 *Immetalia longipalpis* aberr. *bruijni* – STRAND, Lepidopt. Cat., **5**:8.
1912 *Immetalia longipalpis* aberr. *doreyana* – STRAND, Lepidopt. Cat., **5**:8.
1912 *Immetalia longipalpis* aberr. *cynapes* – STRAND, Lepidopt. Cat., **5**:8.

Diagnosis: Bands of both wings white in male; in female, bands also white ("*bruijni*"), or band of fore wings orange yellow and band of hind wings orange red ("*prochyta*"). Band of fore wings short; band of hind wings broader than the black basal area.

Distribution: New Guinea, from Waigioe Island to Astrolabe Bay, southwards to Kapaur.

### Immetalia saturata huonis ROTHSCHILD

1897 *Immetalia meeki* subsp. *huonis* ROTHSCHILD, Novit. Zool., **4**:184 (♂, ♀). – Terra typica: Milne Bay, New Guinea (Types, ♂, ♀, in British Museum).
1912 *Immetalia saturata huonis* – JORDAN in SEITZ, Großschmetterl. Erde, **11**:11.

Diagnosis: Fore wings of both sexes with the band yellowish orange, narrow, reaching the costa; hind wings of both sexes with the band reddish orange, narrower than the basal black area.

Distribution: Gulf of Huon to Milne Bay, New Guinea.

### Immetalia saturata meeki ROTHSCHILD

1896 *Immetalia meeki* ROTHSCHILD, Novit. Zool., **3**:32 (♂). – Terra typica: Louisiade Archipelago (Type, ♂, in British Museum).
1901 *Immetalia meeci* – HAMPSON (ex errore pro *meeki*), Cat. Lepidopt. Phal. Brit. Mus., **3**:536; pl. 51, fig. 16.
1912 *Immetalia saturata meeki* – JORDAN in SEITZ, Großschmetterl. Erde, **11**:11.

Diagnosis: Differs from *I. s. huonis* in the paler bands; band of hind wings somewhat broader.

Distribution: Louisiade Archipelago (D'entrecasteaux, Goodenough, St. Aignan, Fergusson Islands).

### Immetalia saturata diversa ROTHSCHILD & JORDAN

1903 *Immetalia diversa* ROTHSCHILD & JORDAN, Novit. Zool., **10**:486; pl. 12, fig. 24, 25 (♂, ♀). – Terra typica: Aroa River, New Guinea (Types, ♂, ♀, in British Museum).

1912 *Immetalia longipalpis diversa* – JORDAN in SEITZ, Großschmetterl. Erde, **11**:11; pl. 2b.
1912 *Immetalia saturata diversa* forma *leucotaenia* JORDAN in SEITZ, Großschmetterl. Erde, **11**:11; pl. 2b (♂, ♀). – Terra typica: Snow Mts. (Types, ♂, ♀, in British Museum).
1912 *Immetalia saturata diversa* forma *calida* JORDAN in SEITZ, Großschmetterl. Erde, **11**:11 pl. 2b (♂, ♀). – Terra typica: Mts. Owen Stanley, Snow Mts. (Types, ♂, ♀, in British Museum).
1912 *Immetalia saturata diversa* forma *hemigrapha* JORDAN in SEITZ, Großschmetterl. Erde, **11**:11; pl. 2b (♀). – Terra typica: Snow Mts. (Type, ♀, in British Museum).

Diagnosis: Bands of fore wings broader proximally, to the base of vein 2 and encroaching into the cell; breadth and length of band on both wings variable, band of hind wings mostly narrow; short and sometimes absent; colour variable: Bands of both wings in both sexes white ("*leucotaenia*"); band of fore wings white, band of hind wing orange red (*I. s. diversa*); band of fore wings yellowish orange, band of hind wings orange red ("*calida*"); band of fore wings white, band of hind wings absent ("*hemigrapha*").

Distribution: Snow Mountains, Aroa River, New Guinea.

## Immetalia saturata tyrianthina (BUTLER)

1879 *Agarista tyrianthina* BUTLER, Proc. Zool. Soc. London, **1879**:160 (♂). – Terra typica: New Ireland (Type, ♂, in British Museum).
1892 *Phalaenoides tyrianthina* – KIRBY, Syn. Cat. Lepidopt. Het.:23.
1901 *Immetalia longipalpis* – HAMPSON (partim), Cat. Lepidopt. Phal. Brit. Mus., **3**:535.
1912 *Immetalia saturata tyrianthina* – JORDAN in SEITZ, Großschmetterl. Erde, **11**:11.

Diagnosis: Second joint of palpi without white scales; band of fore wings shortened to a spot reaching cell but not into cell, running from vein 5 to vein 2, pointed distally; band of hind wings much broadening, extending to the base, rounded; seventh urite yellow.

Distribution: New Ireland (Terra typica), New Britain.

## Immetalia cyanea ROTHSCHILD

1896 *Immetalia cyanea* ROTHSCHILD, Novit. Zool., **3**:33 (♂, ♀). – Terra typica: Biak Island (Types, ♂, ♀, in British Museum).

Diagnosis: Body and wings black with strong blue reflections; sides of 1st and 2nd joints of palpi, frons above and occiput with some yellowish hair; two last abdominal joints orange red. Fore wings with a silvery blue streak in basal half below costa; a spot in end of cell, and some scales on discocellulars. Length of fore wing 24 mm.

Distribution: Irian (Biak Island, Geelvink Bay).

## Immetalia celebensis ROTHSCHILD

1896 *Immetalia celebensis* ROTHSCHILD, Novit. Zool., **3**:33 (♀). – Terra typica: Celebes, (Type, ♀, in British Museum).

Diagnosis: Body and wings black; basal joint of antennae, sides of frons, ring round neck, and dorsal and subdorsal streaks on thorax above, yellowish white; tegulae fringed with white below, and with a pale yellow lateral patch at base; pectus with yellowish pilosity; femora streaked with yellow, fringed with long and dense hair below. Fore wings with a postmedial orange band beyond cell just below costa to near tornus, with its distal edge slightly angled on vein 4, and its proximal edge waved; oblique silvery blue subbasal and antemedial lines, a bar across middle of cell, and another on discocellulars; cilia white apically and at tornus. Hind wings unspotted; cilia mixed with black and white. Length of fore wing 26 mm.

Distribution: Celebes.

### Immetalia celebensis celebensis ROTHSCHILD

1912 *Immetalia celebensis celebensis* – JORDAN in SEITZ, Großschmetterl. Erde, **11**:11; pl. 2a.

Diagnosis: Cilia of hind wings pure white.

Distribution: South Celebes.

### Immetalia celebensis heydeni (PAGENSTECHER)

1896 *Episteme heydeni* PAGENSTECHER, Entomol. Nachr. Bl., Wien, **22**:50 (♂, ♀). – Terra typica: Northern Celebes (Types, ♂, ♀, in Naturhist. Mus. Wiesbaden).
1901 *Immetalia celebensis* – HAMPSON (partim), Cat. Lepidopt. Phal. Brit. Mus., **3**:537.
1912 *Immetalia celebensis heydeni* – JORDAN in SEITZ, Großschmetterl. Erde, **11**:11.

Diagnosis: Cilia of hind wings mixed black and white.

Distribution: North Celebes.

## Immetalia eichhorni ROTHSCHILD & JORDAN

1901 *Immetalia eichhorni* ROTHSCHILD & JORDAN, Novit. Zool., **8**:406 (♂, ♀). – Terra typica: Isabel Island, Solomon Islands (Type, ♂, in British Museum).

Diagnosis: Body and wings black, the former with strong blue reflections; frons with blue scales; collar edged with blue; legs with blue markings; neck with a distinct white ring. Fore wing with an orange red band, arched, from just below costa at middle to near tornus, touching or entering cell, rounded at its tornal extremity; silvery blue streaks at base; silvery blue subcostal patches at base and before middle, and a cell-spot; cilia white apically. Hind wings unspotted, in male without inner fold or hair-fringe on the lower submedia; cilia tipped with white. Length of fore wing 27 mm.

Distribution: The Solomons: Bougainville, Choiseul, Isabel, Florida Islands.

## Genus **Ophthalmis** HÜBNER

(fig. 12)

1819 *Ophthalmis* HÜBNER, Verz. bek. Schmetterl., 11:166. – Species typica: *Geometra lincea* CRAMER (by subsequent designation: KIRBY 1892).

Diagnosis: Proboscis fully developed; antennae distinctly dilated between the middle and the apex; palpi upturned; 1st joint with pilosity in front; 2nd joint roughly scaled in front; 3rd joint long, naked, slightly porrect; frons with a rounded prominence; pectus and tibiae nearly smoothly scaled; hairtufts at the base of abdomen weakly developed or absent. Venation: in fore wings, vein 2 from about five sevenths of cell; 3 and 4 approximated, about at the same distance from each other as are 4 and 5; discocellulars faintly incurved; 6 from angle of cell; areole narrow; 7 and 8 + 9 from the extremity of areole; 10 from the upper margin of areole, near to the extremity. Hind wings with vein 2 from two thirds of cell; 3, 4 and 6, 7 from the cell-angles; lower discocellular incurved; upper discocellular nearly straight, oblique; 5 obsolescent, from middle of the discocellulars.

Male genitalia: Uncus much elongated, longer than the sacculus plus the termen of valva, moderately slender, faintly but regularly curved, with the terminal hook very slender. Tubus analis narrow, about three quarters of uncus in length, ending in a spoon-shaped process. Vinculum very narrow, proximally with a small bunch of stiff hair. Valva leaf-shaped; costa angled outwards at one third of its length,

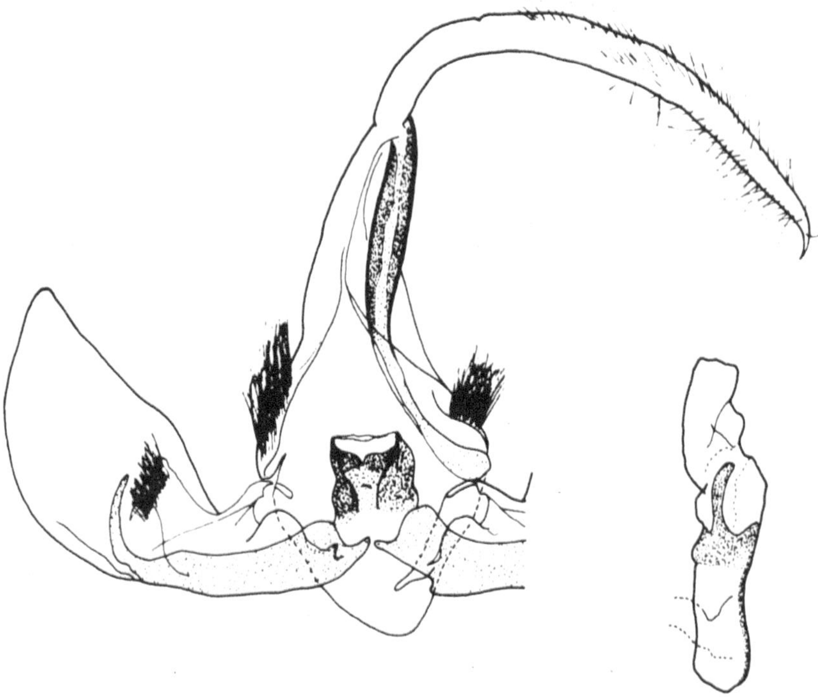

Fig. 12. *Ophthalmis lincea* (CRAMER). B. M. Noct. 8277

then nearly straight; apex slightly upturned, forming an angle of about 30°; sacculus and termen forming together a curve; harpe at the extremity of the rather narrow fold of sacculus, about one-third of costa in length, curved at base, then faintly arched. Aedeagus short, slightly longer than one-half of costa (without the terminal process), stout and straight, with a digitiform terminal process; fultura inferior more or less square, with two ridges beginning at the distal angles and running obliquely inwards. Saccus short, rounded.

Distribution: Philippine Islands, Indonesia, Bismarck Archipelago, Solomon Islands.

Key to the species

| | | |
|---|---|---|
| 1 | Fore wings with an oblique orange band | 2 |
| 1′ | Fore wings without oblique orange band | 3 |
| 2 | Fore wings with an oblique orange antemedial spot | ***O. darna***, p. 69 |
| 2′ | Fore wings without oblique postmedial orange spot | ***O. swinhoei***, p. 69 |
| 3 | Hind wings with orange or yellow markings | 4 |
| 3′ | Hind wings with a postmedial series of whitish spots, or hind wings unmarked | 6 |
| 4 | Hind wings with a postmedial yellow band | ***O. haemorrhoidalis***, p. 70 |
| 4′ | Hind wings with an orange terminal band | 5 |
| 5 | Fore wings with the cilia orange | ***O. cincta***, p. 71 |
| 5′ | Fore wings with the cilia black | ***O. lincea***, p. 71 |
| 6 | Hind wings with a postmedial series of whitish spots | ***O. milete***, p. 74 |
| 6′ | Hind wings black, with only the cilia white | 7 |
| 7 | Fore wings with a white postmedial band | ***O. privata***, p. 77 |
| 7′ | Fore wings without white postmedial band | ***O. floresiana***, p. 77 |

## Ophthalmis darna (DRUCE)

1894 *Agarista darna* DRUCE, Ann. Mag. Nat. Hist., London, (6) **14**:22 (♂). – Terra typica: Timor (Type, ♂, in British Museum).
1901 *Ophthalmis darna* – HAMPSON, Cat. Lepidopt. Phal. Brit. Mus., **3**:645; pl. 54, fig. 11.

Diagnosis: Body and wings blackish brown; palpi with the 2nd joint white in front; frons white above laterally; collar and fore coxae in front orange; anal tuft of abdomen orange at the extremity. Fore wings with an oblique wedge-shaped orange spot from below costa before middle to below the media; a broad orange band from just below costa beyond middle to just below vein 2 near termen, angled inwards below costa, and its distal margin excurved in interval IV; a slight postmedian orange streak in submedian fold; some leaden-silvery marks, viz.: short streaks from base below costa and above vein 1; an oblique streak at cell-extremity; and a discoidal line. Hind wings unmarked, with blue reflections. Length of fore wings 20 mm.

Distribution: Indonesia (former Dutch Timor).

## Ophthalmis swinhoei (SEMPER)

1899 *Scrobigera swinhoei* SEMPER, Reisen Arch. Philippinen, **2**:511; pl. 55, fig. 10 (♂, ♀). – Terra typica: Philippine Islands (Types, ♂, ♀, in Senckenberg Museum).
1901 *Ophthalmis svinhoei* HAMPSON (ex errore pro *swinhoei*), Cat. Lepidopt. Phal. Brit. Mus., **3**:646.
1912 *Ophthalmis swinhoei* – STRAND, Lepidopt. Cat., **5**:38.

Diagnosis: Body and wings black with blue reflections; sides of palpi and of frons and neck with white scales; collar, thorax above and fore coxae orange; male with an orange yellow anal tuft. Fore wings with a broad oblique orange yellow band from just below costa beyond middle to just above tornus where it is rounded; edges slightly indented at the veins; cilia white apically. Hind wings with the basal half with blue reflections; cilia white apically. Length of fore wing 24 – 30 mm.

Distribution: Philippine Islands (East Mindanao).

## **Ophthalmis haemorrhoidalis** (GUÉRIN)

1832 *Agarista haemorrhoidalis* GUÉRIN, Voyage La Coquille, **2**:283; pl. 19, fig. 1. – Terra typica: New Ireland.
1892 *Phalaenoides haemorrhoidalis* – KIRBY, Syn. Cat. Lepidopt. Het.: 22.
1901 *Ophthalmis haemorrhoidalis* – HAMPSON, Cat. Lepidopt. Phal. Brit. Mus., **3**:647; pl. 54, fig. 21 (♀).

Diagnosis: Body and wings blackish brown; palpi with some white scales; 1st joint orange yellow at base; collar with an orange yellow ring, broader beneath; anal tuft of abdomen orange yellow. Fore wings with a yellowish-white subterminal patch with black scales on it between veins 1 and 3, often absent; some blue markings, viz.: scales forming streaks above the subcostal vein and on vein 1; an oblique bar in middle of cell; a discocellular bar; cilia white. Hind wings with a broad yellowish-white band with diffused edges, running from vein 6 to above tornus; cilia white. Length of fore wings 28 mm.

Distribution: Bismarck Archipelago.

### Ophthalmis haemorrhoidalis haemorrhoidalis (GUÉRIN)

Diagnosis: Fore wings with the subtornal spots absent in male, one or two in female; in hind wings, spot in front of vein 5 generally absent in male.

Distribution: New Ireland.

### Ophthalimis haemorrhoidalis gentilis JORDAN

1926 *Ophthalmis haemorrhoidalis* subsp. *gentilis* JORDAN, Novit. Zool., **33**:360 (♂, ♀). – Terra typica: New Hanover (Types, ♂, ♀, in British Museum).

Diagnosis: Fore wings with the subtornal spots double, diffused or single in male, one or two in female; hind wings with the band separated at tornus from the cilia by a black terminal line.

Distribution: New Hanover.

### Ophthalmis haemorrhoidalis castalis JORDAN

1926 *Ophthalmis haemorrhoidalis* subsp. *castalis* JORDAN, Novit. Zool., 33:361 (♂, ♀). – Terra typica: New Britain (Types, ♂, ♀, in British Museum).

Diagnosis: Hind wings with the subtornal spots present in both sexes; fore spot nearer to termen in female; hind wings with the pale band broadly united with the cilia at tornus, and extending further forward.

Distribution: New Britain.

## Ophthalmis cincta (BOISDUVAL)

1874 *Agarista cincta* BOISDUVAL, Rev. Mag. Zool., (3) 2:80. – Terra typica: Gilolo (Type, ♂, in British Museum).
1887 *Ophthalmis zelleri* BUTLER, Ann. Mag. Nat. Hist., London, (5) 19:215 (♂). – Terra typica: Ternate (Type, ♂, in British Museum).
1901 *Ophthalmis cincta* – HAMPSON, Cat. Lepidopt. Phal. Brit. Mus., 3:647; pl. 54, fig. 22 (♂).

Diagnosis: Body and wings blackish brown; collar, fore coxae and the two last urites orange. Fore wings with a terminal orange patch and orange termen; costal area sometimes orange; some silvery-blue markings, viz.: streaks below costa and on vein 1; an oblique streak in cell; another below cell; a discocellular bar; veins beyond the cell with silvery blue streaks; cilia orange. Hind wings with a rather broad orange terminal band; cilia orange. Length of fore wing 24 mm.

Distribution: Moluccas and some neighbouring islands.

### Ophthalmis cincta cincta (BOISDUVAL)

1912 *Ophthalmis cincta cincta* – JORDAN in SEITZ, Großschmetterl. Erde, 11:19; pl. 3g.

Diagnosis: Fore wings without orange costal streak.

Distribution: Halmaheira, Ternate, Batjan, Obi Islands.

### Ophthalmis cincta proxanthia HAMPSON

1901 *Ophthalmis proxanthia* HAMPSON, Cat. Lepidopt. Phal. Brit. Mus., 3:647; pl. 54, fig. 5 (♀). – Terra typica: Mount Mada, Buru Island (Type, ♀, in British Museum).
1912 *Ophthalmis cincta proxanthia* – JORDAN in SEITZ, Großschmetterl. Erde, 11:19.

Diagnosis: Fore wings with the costal area orange.

Distribution: Boeroe.

## Ophthalmis lincea (CRAMER)

1779 *Geometra lincea* CRAMER, Pap. Exot., 3: pl. 228 B. – Terra typica: "Suriname" (ex errore) = Amboina (BOISDUVAL 1874); type not designated.
1819 *Ophthalmis lincea* – HÜBNER, Verzeichnis bek. Schmetterl., (11) 00:166.

D i a g n o s i s : Body and wings black; ring around the neck, sometimes the whole collar, fore coxae and the two last urites orange. Fore wings with an apical orange patch, very variable in size; some silvery blue markings, variable in extent, viz.: a streak below the basal half of costa; a streak on vein 1; an oblique streak in cell-middle; a short streak below middle of cell; a discocellular lunule; a postmedial row of nervural streaks. Hind wings with blue reflections; an orange or yellow orange terminal band, varying in width. Fore wings averaging 25 mm in length.

D i s t r i b u t i o n : The Philippine Islands to the Solomons, apparently absent on Halmaheira, Ternate, Batjan and Obi, where it is replaced by *O. cincta*.

### Ophthalmis lincea lincea (CRAMER)

1856 *Lithosia formosa* MONTROUZIER, Ann. Sci. Phys. Nat., Lyon, (2) 8:410. – T e r r a t y p i c a : East Indies (Type not designated).
1874 *Agarista linceoides* BOISDUVAL, Rev. Mag. Zool., (3) 2:80. – T e r r a t y p i c a : New Guinea (Type in British Museum).
1892 *Ophthalmis formosa* – KIRBY, Syn. Cat. Lepidopt. Het.:31.
1911 *Ophthalmis lincea* var. *leucapex* STRAND, Mitt. Zool. Mus. Berlin, 5:474 (♂, ♀). – T e r r a t y p i c a : German Neuguinea = Papua (Type in Berlin Museum).
1912 *Ophthalmis lincea lincea* – JORDAN in SEITZ, Großschmetterl. Erde, 11:19.

D i a g n o s i s : Hind wings with the orange band extending to middle of the wing, with the proximal margin rather straight; sometimes, space between cell-tip and costa sprinkled with orange.

D i s t r i b u t i o n : The Moluccas: Boeroe, Amboina, Saparoea, Goram-Laoet, Watubela, Manovolka, Groot and Klein Key, Aroe Islands, Irian, Papua, Louisiade Archipelago, Woodlark Island.

### Ophthalmis lincea bambucina (ESCHSCHOLTZ)

1830 *Noctua bambucina* ESCHSCHOLTZ, Kotzebues Entdeckungsreise, 3:219; pl. 11, fig. 30. – T e r r a t y p i c a : Philippine Islands.
1892 *Ophthalmis bambucina* – KIRBY, Syn. Cat. Lepidopt. Het.:31.
1912 *Ophthalmis lincea bambucina* – JORDAN in SEITZ, Großschmetterl. Erde, 11:19.

D i a g n o s i s : Hind wings with the band yellow-orange, narrow, especially apically and tornally, with the proximal margin parallel with the termen.

D i s t r i b u t i o n : Philippine Islands.

### Ophthalmis lincea intermedia JORDAN

1912 *Ophthalmis lincea intermedia* JORDAN in SEITZ, Großschmetterl. Erde, 11:19 (♂). – T e r r a t y p i c a : Palawan (Type, ♂, in British Museum).

D i a g n o s i s : Hind wings with the band as narrow as in *O. l. bambucina*, but deep orange; blue markings well developed.

D i s t r i b u t i o n : Palawan Island, Cagayan Soeloe, Balabak, Sangir.

## Ophthalmis lincea bismarcki JORDAN

1912 *Ophthalmis lincea bismarcki* JORDAN in SEITZ, Großschmetterl. Erde, **11**:19 (♂). – Terra typica: New Hanover (Type, ♂, in British Museum).

Diagnosis: Fore wings with the apical spot very much reduced, hardly more than vestigial; blue markings well developed. Hind wings as in *O. l. lincea*.

Distribution: Bismarck Archipelago: New Hanover (Terra typica), New Britain, Duke of York Island.

## Ophthalmis lincea aluensis BUTLER

1887 *Ophthalmis aluensis* BUTLER, Ann. Mag. Nat. Hist., London, (5) **19**:215 (♂). – Terra typica: Alua, Solomon Islands (Type in British Museum).
1912 *Ophthalmis lincea aluensis* JORDAN in SEITZ, Großschmetterl. Erde, **19**:19.

Diagnosis: Fore wings with the apical spot about as in *O. l. bismarcki*, in female often as extensive as in *O. l. lincea*; hind wings with the terminal band variable, but always with the proximal margin parallel with the termen.

Distribution: Solomon Islands.

## Ophthalmis lincea dimidiata JORDAN

1912 *Ophthalmis lincea dimidiata* JORDAN in SEITZ, Großschmetterl. Erde, **11**:19 (♂). – Terra typica: Biak Island (Type, ♂, in British Museum).
1912 *Ophthalmis lincea* var. *biakensis* STRAND, Lepidopt. Cat., **5**:37. – Terra typica: Biak Island.

Diagnosis: Fore wings altogether without the orange apical patch; hind wings with the terminal band deep orange and very broad.

Distribution: Biak Island.

## Ophthalmis lincea hemixantha JORDAN

1912 *Ophthalmis lincea hemixantha* JORDAN in SEITZ, Großschmetterl. Erde, **11**:19 (♂, ♀). – Terra typica: Tenimber Island (Type, ♂, in British Museum).
1920 *Ophthalmis lincea* subsp. *hemixantha* – HAMPSON, Cat. Lepidopt. Phal. Brit. Mus., Suppl. **2**:590.

Diagnosis: Fore wings with the apical spot large; blue markings weakly developed; hind wings with the terminal band very wide, reaching to the lower angle of cell; colour intensity intermediate between *O. l. lincea* and *O. l. bambucina*.

Distribution: Tenimber Island.

## Ophthalmis lincea admiralitatis ROTHSCHILD

1916 *Ophthalmis lincea* subsp. *admiralitatis* ROTHSCHILD, Novit. Zool., **23**:334 (♂, ♀). – Terra typica: Admiralty Island (Types, ♂, ♀, in British Museum).

Diagnosis: Fore wings without the orange apical patch; cilia white apically; hind wings with the terminal half deeply orange yellow.

Distribution: Admiralty Island, Manus Island.

## Ophthalmis milete (CRAMER)

1775 *Noctua milete* CRAMER, Pap. Exot., **1**:pl. 18 D. – Terra typica: East Indies. – Terra typica restricta: Sumatra [JORDAN 1912] (Type not designated).

Diagnosis: Body and wings black; palpi streaked with white laterally; frontal tuft white above; collar, throat and coxae orange; tegulae edged with white above; metathorax with a white spot; legs yellowish white; abdomen with yellowish white segmental bands; extreme tip and underside orange. Fore wings with the costal edge orange to beyond the middle in male; markings silvery blue and yellowish white or orange, viz.: subbasal spots below costa and cell; a streak on basal half of vein 1; an antemedial subcostal spot; a bar from cell to the streak on vein 1; a spot below middle of costa; an oblique bar across middle of cell, sometimes bent inwards to the origin of vein 2 and continued to vein 1; obliquely placed white spots in a below middle of cell; an oblique bar at the cell-end; a postmedial series of elongate spots, those above veins 2, 3 and 4 more distal, that below vein 2 more proximal; a more or less developed subterminal series of silvery blue dots; cilia white apically and tornally. Hind wings with whitish spots in and below middle of cell; a silvery blue discocellular bar; a curved postmedial series of whitish spots, spot below vein 2 more proximal; a more or less developed subterminal series of silvery blue spots; cilia mostly white. Length of fore wing 21 – 28 mm.

Early stages: Larva dark slate colour, with subdorsal and lateral series of black spots; pilosity sparse; head and prolegs yellow or orange. On *Cissus*.

Distribution: Sumatra to Soemba and Celebes.

## Ophthalmis milete milete (CRAMER)

1864 *Eusemia conferta* WALKER, List Lepidopt. Het. Brit. Mus., **31**:53 (♀). – Terra typica: Sumatra (Types, ♂, ♀, in Museum Oxford University).
1892 *Phalaenoides confertus* – KIRBY, Syn. Cat. Lepidopt. Het.: 23.
1912 *Ophthalmis milete milete* – JORDAN in SEITZ, Großschmetterl. Erde, **11**:19; pl. 3h.

Diagnosis: Blue markings well developed; abdomen, underside in female orange, banded with black.

Distribution: Sumatra; possibly Engano Island.

## Ophthalmis milete leucisca JORDAN

1912 *Ophthalmis milete leucisca* JORDAN in SEITZ, Großschmetterl. Erde, **11**:19 (♂, ♀). – Terra typica: Java (Types, ♂, ♀, in British Museum).
1912 *Ophthalmis milete leucisca* aberr. *niveata* – JORDAN in SEITZ, Großschmetterl. Erde, **11**:19.

Diagnosis: Abdomen as in *O. m. milete*, with more of black on the last urite in female; white spots on wings larger; blue markings reduced, especially the discal streaks.

Distribution: Java, Bali, Bawean Island.

## Ophthalmis milete torva JORDAN

1912 *Ophthalmis milete torva* JORDAN in SEITZ, Großschmetterl. Erde, **11**:19–20 (♂, ♀). – Terra typica: Lombok (Types, ♂, ♀, in British Museum).

Diagnosis: Abdomen with more of black; light rings broader; underside with urites II–VI with the light margins obsolescent; both wings with the white spots reduced, at least the proximal ones; discal spots pure white on fore wings; blue markings better developed than in *O. m. leucisca*.

Distribution: Lombok Island.

## Ophthalmis milete ornata JORDAN

1912 *Ophthalmis milete ornata* JORDAN in SEITZ, Großschmetterl. Erde, **11**:20 (♂, ♀). – Terra typica: Sumbawa (Types, ♂, ♀, in British Museum).

Diagnosis: Abdomen blacker, with the pale margins narrower; half of urite VII black dorsally in female; underside with much less orange; wings with the blue markings as in *O. m. milete*, but the white proximal spots smaller, the distal white spots larger.

Distribution: Soembawa Island.

## Ophthalmis milete sumbana JORDAN

1912 *Ophthalmis milete sumbana* JORDAN in SEITZ, Großschmetterl. Erde, **11**:20. – Terra typica: Sumba (Type, ♂, in British Museum).

Diagnosis: Differs from *O. m. torva* in that the white proximal spots are smaller; the discal spots on hind wings slightly more proximal; blue markings less well developed.

Distribution: Soemba Island.

## Ophthalmis milete mutata (WALKER)

1864 *Eusemia mutata* WALKER, List Lepidopt. Het. Brit. Mus., **31**:52 (♀). – Terra typica: Celebes (Type, ♀, in Museum Oxford University).
1864 *Eusemia fenestrata* WALKER, List. Lepidopt. Het. Brit. Mus., **31**:54 (♂). – Terra typica: Menado (Type, ♂, in Museum Oxford University).
1874 *Agarista rosenbergi* FELDER, Reise Novara: pl. 107, fig. 1. – Terra typica: Celebes (Type in British Museum).
1874 *Eusemia rosenbergi* – BOISDUVAL, Rev. Mag. Zool., (3) **2**:94.
1892 *Phalaenoides mutatus* – KIRBY, Syn. Cat. Lepidopt. Het.: 24.
1892 *Phalaenoides fenestratus* – KIRBY, Syn. Cat. Lepidopt. Het.: 23.
1912 *Ophthalmis milete mutata* – JORDAN in SEITZ, Großschmetterl. Erde, **11**:20.

Diagnosis: Larger than the other races; antennae with with whitish scales.

Distribution: Northern, Eastern, and Southern Celebes.

## Ophthalmis milete boetonensis JURIAANSE

1920 *Ophthalmis milete boetonensis* JURIAANSE, Tijdschr. Entomol., 's Gravenhage, Amsterdam, **62**:49. – Terra typica: Butung Island (Type in Museum Amsterdam).

Diagnosis: Markings orange.

Distribution: Butung Island.

## Ophthalmis milete roeberi (RIBBE)

1889 *Eusemia roeberi* RIBBE, Dtsch. entomol. Z. Iris, Dresden, **2**:184; pl. 3, fig. 3 (♂). – Terra typica: Banggaja Island.
1892 *Phalaenoides roeberi* – KIRBY, Syn. Cat. Lepidopt. Het.: 24.
1912 *Ophthalmis milete roeberi* – JORDAN in SEITZ, Großschmetterl. Erde, **11**:20.

Diagnosis: Antennae white only subterminally; both wings with the markings slightly larger than in *O. m. mutata*; in fore wings, space between the white spots before submedia orbicular.

Distribution: Banggaja Island, east of Celebes.

## Ophthalmis milete basalis ROTHSCHILD

1896 *Ophthalmis basalis* ROTHSCHILD, Novit. Zool., **3**:50; pl. 14, fig. 25 (♀). – Terra typica: Sula-Mangoli Island (Type, ♀, in British Museum).

Diagnosis: Abdomen with whitish margins vestigial; underside with large orange spots; antennae white subterminally; fore wings with the medial spots large; hind wings with the basal area white to near cell-extremity and on anal area well beyond middle.

Distribution: Soela-Mangoli.

## Ophthalmis privata (WALKER)

1864 *Agarista privata* WALKER, List Lepidopt. Het. Brit. Mus., 31:48 (♂) – Terra typica: Ceram (Type, ♂, in Museum Oxford University).
1886 *Agarista pagenstecheri* RÖBER, Dtsch. entomol. Z. Iris, Dresden, 1:40; pl. 2, fig. 10. – Terra typica: Ceram.
1892 *Episteme pagenstecheri* – KIRBY, Syn. Cat. Lepidopt. Het.:28.
1892 *Phalaenoides privatus* – KIRBY, Syn. Cat. Lepidopt. Het.:23.
1901 *Ophthalmis privata* – HAMPSON, Cat. Lepidopt. Phal. Brit. Mus., 3:650; pl. 54, fig. 2.

Diagnosis: Body and wings blackish brown; palpi with the first and part of the second joints orange; neck with an orange ring; fore and mid legs slightly streaked with orange; anal tuft orange yellow. Fore wings with a white spot at the upper cell-tip; a white curved band beyond cell between veins 2 and 7, with waved proximal margin; some silvery blue markings, viz.: a few scales forming a streak below the basal half of costa; a spot in middle of cell; another below cell; and a discocellular band; cilia white near tornus. Hind wings unmarked; cilia white. Length of fore wing 26 mm.

Distribution: Ceram (Terra typica), Amboina.

## Ophthalmis floresiana (ROTHSCHILD)

1897 *Crinala floresiana* ROTHSCHILD, Novit. Zool., 4:312 (♂). – Terra typica: Flores Island (Type, ♂, in British Museum).
1901 *Ophthalmis floresiana* – HAMPSON, Cat. Lepidopt. Phal. Brit. Mus., 3:650; pl. 54, fig. 19.

Diagnosis: Body and wings black; abdomen with blue reflections; 2nd joint of palpi white in front; sides of frons spotted with white; neck with a deep orange ring, interrupted in female; fore coxae deep orange; mid tibiae streaked with whitish; anal tuft deep orange. Fore wings with a silvery blue oblique antemedial line; a bar across end of cell; a discocellular bar; a postmedial spot on vein 1; termen and cilia white apically. Hind wings with strong blue reflections; extreme termen and cilia white. Length of fore wing 26 mm.

Distribution: Flores Island.

## Genus Crinala JORDAN

(fig. 13)

1896 *Crinala* JORDAN, Novit. Zool. 3:35. – Species typica: *Crinala mimetica* ROTHSCHILD.

Diagnosis: Proboscis fully developed; antennae dilated towards extremity; palpi upturned, reaching vertex; 1st and 2nd joints with long pilosity; 3rd minute, nearly naked; frons rounded, without prominence; mid and hind tibiae fringed with rough hair above. Venation: in fore wing, vein 2 from 3/5 of cell; 3, 4 and 5 much approximated, nearly equidistant; 6 from below angle of cell; areole small; 7, 10 8 + 9 from extremity of areole. In hind wing, vein 2 from 3/4 of cell; 3, 4 and 6, 7 from angles of cell; discocellulars incurved; 5 obsolescent, from middle of discocellulars. In male, upperside of hind wings bearing long hair between the veins before the termen.

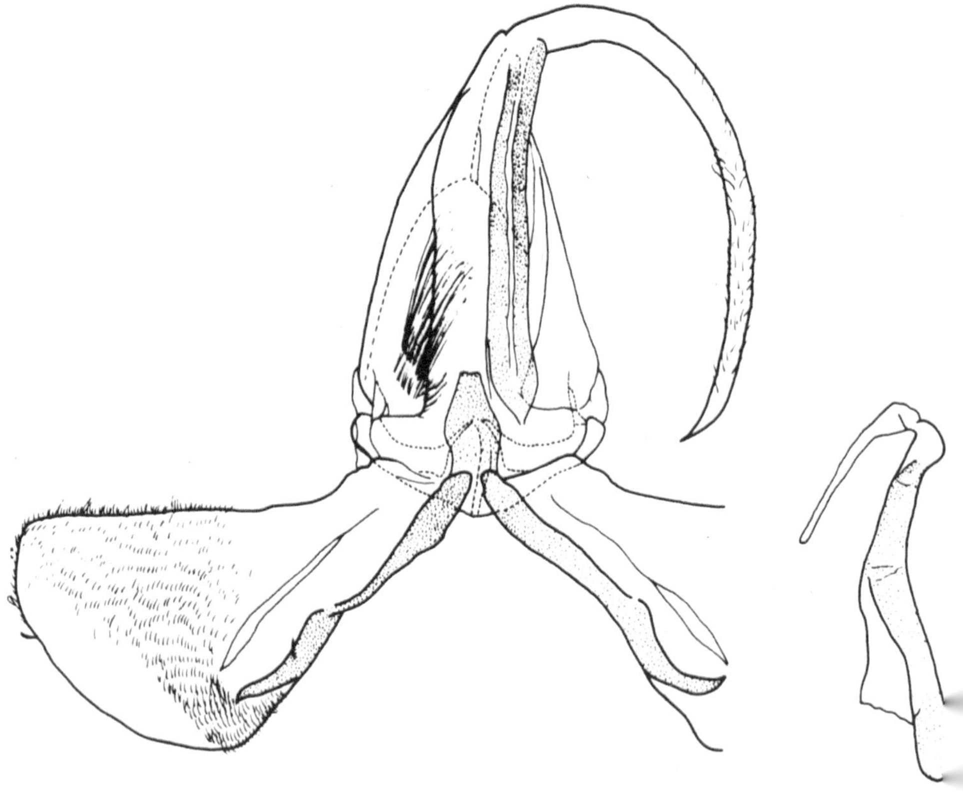

Fig. 13. *Crinala mimetica* ROTHSCHILD. Type. B. M. Noct. 8375

Male genitalia: Uncus long and slender, distinctly longer than the costa, arched and ending in a small hook. Tubus analis about two thirds of uncus in length, narrow, faintly broader terminally. Vinculum without a median production, with a proximal bunch of moderately long, stiff hair. Valva roughly triangular; base narrow, broadening beyond $^1/_4$; costa nearly straight; apex obtuse, rounded; termen evenly rounded; sacculus nearly straight; fold of sacculus narrow, reaching to middle and ending in a short, faintly arched harpe; marginal pilosity rather stiff in distal half of costa, around the apex and in distal portion of sacculus. Aedeagus as long as costa, slender, faintly elbowed at two-thirds, otherwise nearly straight, bearing a slender terminal process, directed obliquely proximad; fultura inferior lance-shaped, but blunt terminally. Saccus hardly developed.

D i s t r i b u t i o n : Philippine Islands.

### **Crinala mimetica** ROTHSCHILD

1896 *Crinala mimetica* ROTHSCHILD, Novit. Zool., 3:36; pl. 14, fig. 24 (♂). – Terra typica: Northern Luzon, Philippine Islands (Type, ♂, in British Museum).

Diagnosis: Head black, mixed with orange hair; antennae black; palpi orange with the third joint black; occiput and collar orange; thorax and abdomen above bluish black; coxae, fore and mid femora and pilosity of fore tibiae orange; abdomen orange on underside except base and extremity. Fore wings bluish black; a silvery blue spot in cell-extremity; spot on discocellulars silvery blue; veins streaked with white from the cell for about half their length; cilia white. Hind wings bluish black; veins very faintly streaked with white in the submarginal area; cilia white. Length of fore wing 26 mm.

Distribution: Philippine Islands: North Luzon.

## Genus **Damias** WALKER

(fig. 14)

1854 *Damias* WALKER, List Lepidopt. Het. Brit. Mus., 1:14. – Species typica: *Damias varia* WALKER (by monotypy).
1864 *Burgena* WALKER, List Lepidopt. Het. Brit. Mus., 31:55. – Species typica: *Burgena educta* WALKER = *Damias varia* WALKER.

Diagnosis: Proboscis fully developed; antennae thick, dilated from middle to extremity; laterally of antennae, long piliform scales directed outwards; palpi upturned, with the first and the second joints bearing moderately long hair in front, and the third joint long, naked, somewhat porrect; eyes naked; frons broad, convex, without conical prominence, but with a raised rim; pectus and femora with long pilosity; tibiae and tarsi with appressed scales; first abdominal ring with the tergite rounded and protruding forwards laterally; urites II–VI bearing a spot of brownish modified scales. Fore wings narrow; venation: vein 2 from three-quarters of cell; 3, 4, 5 much approximated; 6 from upper cell-angle; areole very narrow; 7 and the very short stalk of 10, 8 + 9 from extremity of areole. In hind wings, vein 2 from five-sixths of cell; 3, 4 and 6, 7 from the cell-angles, 3, 4 sometimes short stalked; discocellulars incurved; 5 obsolescent, from middle of discocellulars.

Male genitalia: Uncus as long as the straight portion of sacculus, very slender, ending in a minute hook. Tubus analis slender. Vinculum narrow, with the fold relatively very narrow, but with a proximal subbasal expansion, bearing a small tuft of rather short hair. Valva roughly V-shaped; basal portion straight, moderately broad; terminal portion bent caudad, egg-shaped, at a right angle with the basal part; fold of sacculus very narrow, reaching to the end of the straight portion of valva and ending in a small hook-shaped harpe. Aedeagus about one-quarter longer than the uncus, its distal half formed by a slender, claw-like process, with a few minute distal indentations; fultura inferior semi-elliptical, with the distal margin faintly incurved. Saccus not developed.

Distribution: Moluccas to New Guinea, Australia, the Bismarck Archipelago, the Solomon Islands.

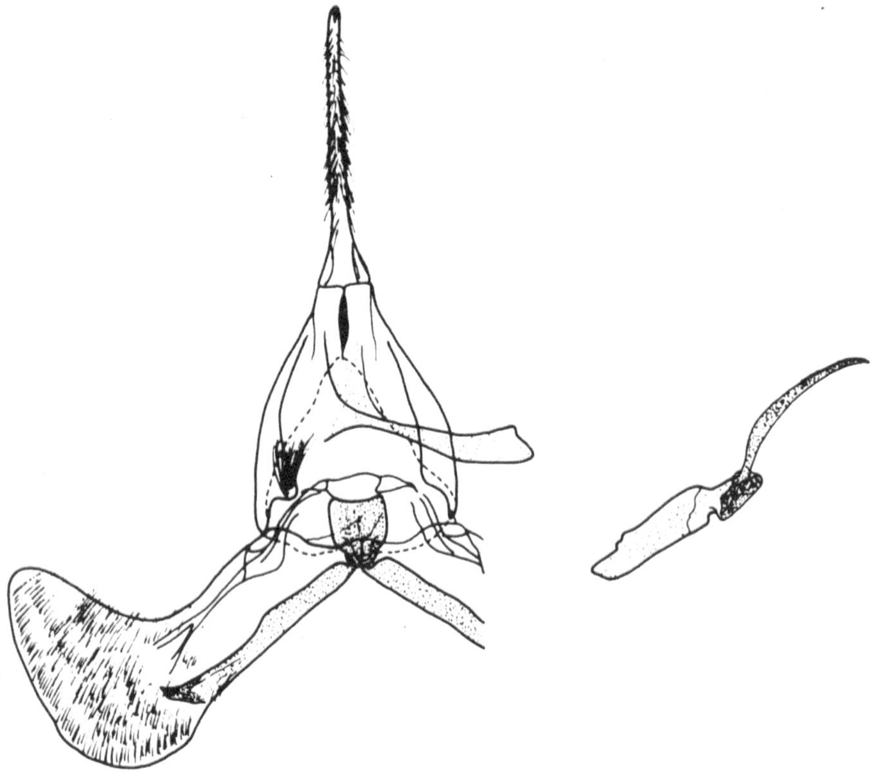

Fig. 14. *Damias varia* WALKER. B. M. Noct. 8390

## Key to the species

1 Hind wings with an orange band . . . . . . . . . . . . . . . . . . . . . . . . . . . . . . 2
1' Hind wings without any band, uniformly black with blue or green reflections . . . . . . *D. chalybeata*, p. 84
2 Fore wings with a whitish or orange streak below cell . . . . . . . . . . . . . . . . . . . . 3
2' Fore wings without white or orange band below cell . . . . . . . . . . *D. splendida*, p. 83
3 Mesothorax with transverse bands . . . . . . . . . . . . . . . . . . . . . . . . . *D. varia*, p. 80
3' Mesothorax with longitudinal stripes . . . . . . . . . . . . . . . . . . . . . *D. constricta*, p. 83

## Damias varia WALKER

1854 *Damias varia* WALKER, List Lepidopt. Het. Brit. Mus., 1:15 (♂). – Terra typica: Richmond River, New South Wales (Type, ♂, in British Museum).
1901 *Burgena varia* – HAMPSON, Cat. Lepidopt. Phal. Brit. Mus., 3:538; fig. 238.

Diagnosis: Head, thorax and wings black; 1st and 2nd palpal joints yellowish in front; sides of frons yellowish white; in front and behind collar and across mesothorax, orange bands; pectus, coxae, femora with orange pilosity; abdomen orange with dorsal and lateral black bands, those on IV and V urites dorsally confluent in male; underside orange. Fore wings with a short orange or pale yellow fascia below base of cell; a pale yellow square spot in cell-middle; a spot in interspace VI, and mostly conjoined spots in interspace III and IV; cilia sometimes white apically. Hind wings with a broad orange medial band from above vein 6 to the anal margin, more or less dilated distally between veins 3 and 6.

Distribution: Moluccas through New Guinea to Australia.

### Damias varia varia WALKER

1912 *Damias varia varia* – JORDAN in SEITZ, Großschmetterl. Erde, **11**:12; pl. 2b.

Diagnosis: As above; basal streak on fore wings pale yellow; cilia not white apically.

Distribution: Northern New South Wales, Queensland.

### Damias varia euxanthe JORDAN

1912 *Damias varia euxanthe* JORDAN in SEITZ, Großschmetterl. Erde, **11**:12. – Terra typica: St. Aignan Island, Louisiade Archipelago (Type, ♂, in British Museum).

Diagnosis: Fore wings generally with a small bluish white spot at base of cell; postdiscal spots in interspaces III and IV separated; cilia white apically; hind wings with the orange band very wide.

Distribution: Louisiade Archipelago (St. Aignan Island).

### Damias varia anisa JORDAN

1912 *Damias varia anisa* JORDAN in SEITZ, Großschmetterl. Erde, **11**:12. – Terra typica: Kiriwimi Island, Trobriand Group (Type, ♂, in British Museum).

Diagnosis: Fore wings with the spot in cell-middle small, transversal; spots in interspaces III and IV separate, lower spot short; hind wings with the orange band as broad as the black marginal space; cilia white apically.

Distribution: Trobriand Group (Kiriwimi Island).

### Damias varia ravida JORDAN

1912 *Damias varia ravida* JORDAN in SEITZ, Großschmetterl. Erde, **11**:12 (♂, ♀). – Terra typica: Fergusson and Goodenough Islands, D'Entrecasteaux Islands (Types, ♂, ♀, in British Museum).

Diagnosis: Fore wings with spots in interspaces III and IV separate, at least twice as large as the spot in cell; lower spot generally cut obliquely; hind wings as in *D. v. anisa*.

Distribution: D'Entrecasteaux Islands (Fergusson and Goodenough Islands).

### Damias varia baia JORDAN

1912 *Damias varia baia* JORDAN in SEITZ, Großschmetterl. Erde, **11**:12 (♂, ♀). – Terra typica: Mountains of Papua (Types, ♂, ♀, in British Museum).

Diagnosis: Fore wings with the spots small; cilia not white apically.

Distribution: Mountains of Papua, New Guinea.

### Damias varia educta (WALKER)

1864 *Burgena educta* WALKER, List Lepidopt. Het. Brit. Mus., **31**:56 (♀). – Terra typica: Kei Island (Type, ♀, in British Museum).
1874 *Eusemia arruana* BOISDUVAL, Rev. Mag. Zool., (3) **2**:100. – Terra typica: "Arrow" = Aru Island (Type, ♂, in British Museum).
1892 *Episteme arruana* – KIRBY, Syn. Cat. Lepidopt. Het.:28.
1912 *Damias varia educta* – JORDAN in SEITZ, Großschmetterl. Erde, **11**:12.

Diagnosis: Fore wings with the postdiscal spots in interspaces III and IV square, rarely more or less circular; cilia with a few white scales at the apex.

Distribution: Kei Islands (Terra typica), Aru Islands, Waigiu, islands of Geelvink Bay, New Guinea, at lower and middle altitudes.

### Damias varia transducta (WALKER)

1864 *Burgena transducta* WALKER, List Lepidopt. Het. Brit. Mus., **31**:56 (♂). – Terra typica: Gilolo Island (Type, ♂, in British Museum).
1912 *Damias varia transducta* – JORDAN in SEITZ, Großschmetterl. Erde, **11**:13.
1912 *Damias varia* aberr. *transducta* – STRAND, Lepidopt. Cat., **5**:9.

Diagnosis: Fore wings with the postdiscal spots in interspaces III and IV large, separated in three; hind wings with the orange band strongly narrowed at middle.

Distribution: Northern Moluccas (Halmaheira, Batjan, Ternate, Morotai, Gilolo).

### Damias varia tripartita JOICEY & TALBOT

1916 *Damias varia tripartita* JOICEY & TALBOT, Trans. Entomol. Soc. London, **1916**:80 (♂, ♀). – Terra typica: Schouten Island (Types, ♂, ♀, in British Museum).

D i a g n o s i s: Fore wings with the postmedian pale spot trifid as in *D. v. transducta*, but smaller, extending to interspace V; cell-spot minute in male, obsolete in female; in hind wings, the dark margin wider; cilia entirely black.

D i s t r i b u t i o n: Irian (Schouten Island).

### **Damias constricta** (ROTHSCHILD & JORDAN)

1905 *Burgena constricta* ROTHSCHILD & JORDAN, Novit. Zool., **12**:472 (♀). – T e r r a t y p i c a: New Guinea (Type, ♀, in British Museum).
1912 *Damias constricta* – JORDAN in SEITZ, Großschmetterl. Erde, **11**:13; pl. 2b.

D i a g n o s i s: Head, thorax and wings black; mesonotum yellow with black stripes; abdomen yellow, banded with black; last urite yellow. Fore wings with a basal yellow streak running below cell and along the submedian vein, more elongate than in *D. varia*; cell with a subbasal whitish yellow spot; rest as in *D. varia*. Hind wings with the yellow band strangled at middle, not encroaching in cell and reaching vein 2 in front.

D i s t r i b u t i o n: Papua: Mounts Owen-Stanley at about 2000 m altidude.

### **Damias splendida** (BUTLER)

1887 *Eusemia splendida* BUTLER, Ann. Mag. Nat. Hist., London, (5) **19**:214 (♂). – T e r r a t y p i c a: Guadalcanal, Solomon Islands (Type, ♂, in British Museum).
1892 *Episteme splendida* – KIRBY, Syn. Cat. Lepidopt. Het.:31.
1901 *Burgena splendida* – HAMPSON, Cat. Lepidopt. Phal. Brit. Mus., **3**:539; pl. 51, fig. 14.
1912 *Damias splendida* – JORDAN in SEITZ, Großschmetterl. Erde, **11**:13.

D i a g n o s i s: Head, thorax and wings black with blue reflections; palpi mixed with white in front at the base; frons white at the sides; collar, front of pectus, pilosity of coxae (sometimes only of the fore coxae) orange; abdomen orange, with complete black bands from the IV to last urite which is entirely black. Fore wings with a small white antemedial cell-spot; a square spot in end of cell; elongate spots in interspaces II, III and VI; cilia white apically. Hind wings with a broad medial orange band from vein 6 to the anal margin, angled outwards on vein 5. Length of fore wing 27 mm.

D i s t r i b u t i o n: Solomon Islands.

#### D a m i a s s p l e n d i d a s p l e n d i d a (BUTLER)

1912 *Damias splendida splendida* – JORDAN in SEITZ, Großschmetterl. Erde, **11**:13.

D i a g n o s i s: Fore, mid and hind coxae with orange pilosity; hind wings with the orange band broad in both sexes.

D i s t r i b u t i o n: Guadalcanal.

### Damias splendida pectoralis (JORDAN)

1904 *Burgena spelndida* subsp. *pectoralis* JORDAN, Novit. Zool., **11**:444 (♂, ♀). – Terra typica: Solomon Islands (Types, ♂, ♀, in British Museum).
1912 *Damias splendida pectoralis* – JORDAN in SEITZ, Großschmetterl. Erde, **11**:13.

Diagnosis: Palpi with only a few white scales; only the fore coxae with orange pilosity at sides and behind; hind wings with the orange band broad in both sexes.

Distribution: Florida, Santa Isabel Island, Choiseul Island.

### Damias splendida dispar JORDAN

1912 *Damias splendida dispar* JORDAN in SEITZ, Großschmetterl. Erde, **11**:13 (♂, ♀). – Terra typica: Bougainville Island (Types, ♂, ♀, in British Museum).
1920 *Burgena reducta* – HAMPSON (nec JORDAN), Cat. Lepidopt. Phal. Brit. Mus., Suppl. **2**:540.

Diagnosis: Coxae as in *D. v. pectoralis*; yellow bands on urites V and VI very narrow, broadly interrupted; hind wings with the orange band in male much reduced, sometimes a mere spot in the anal area; sometimes obsolescent, orange red; in female well formed, but not broader than half of the terminal black area.

Distribution: Bougainville, Treasury Island.

### Damias splendida reducta (JORDAN)

1903 *Burgene reducta* JORDAN, Novit. Zool., **10**:487 (♂). – Terra typica: Kolombangara, Solomon Islands (Type, ♂, in British Museum).
1912 *Damias splendida reducta* – JORDAN in SEITZ, Großschmetterl. Erde, **11**:13; pl. 2b.

Diagnosis: Hind wings with the orange band strongly reduced in both sexes, like in male of *D. s. dispar*.

Distribution: Kolombangara, Rubian Island.

### **Damias chalybeata** (ROTHSCHILD)

1896 *Burgena chalybeata* ROTHSCHILD, Novit. Zool., **3**:39 (♂). – Terra typica: New Britain (Type, ♂, in British Museum).
1896 *Burgena amoena* ROTHSCHILD, Novit. Zool., **3**:39 (♀). – Terra typica: New Britain (Type, ♀, in British Museum).
1912 *Damias chalybeata* – JORDAN in SEITZ, Großschmetterl. Erde, **11**:13.

Diagnosis: Head, body and wings black with blue reflections; basal antennal joint white; 1st and 2nd palpal joints white at sides; frons with the sides white; coxae and femora fringed with orange hair. Fore wings in male with a broad diffused bluish white subbasal band; in female with a bluish spot in cell, sometimes absent, and a broad postcellular band between veins 1 and 7, bluish towards its extremities, white edged with blue between veins 2 and 5. Hind wings unmarked. Length of fore wing 23 mm.

D i s t r i b u t i o n: Bismarck Archipelago.

### Damias chalybeata chalybeata (ROTHSCHILD)

1896 *Burgena amoena* ROTHSCHILD, Novit. Zool., **3**:39. – T e r r a t y p i c a: New Britain (Type, ♀, in British Museum).
1912 *Damias chalybeata chalybeata* – JORDAN in SEITZ, Großschmetterl. Erde, **11**:13; pl. 2d.

D i a g n o s i s: Pectus, coxae and long hair of femora orange; fore wings in female with a cell-spot; hind wings with the band strongly narrowed and bearing three or four white spots. Length of fore wing 18 – 26 mm.

D i s t r i b u t i o n: New Britain.

### Damias chalybeata leucidia JORDAN

1912 *Damias chalybeata leucidia* JORDAN in SEITZ, Großschmetterl. Erde, **11**:13 (♀). – T e r r a t y p i c a: New Hanover (Type, ♀, in British Museum).
1920 *Burgena chalybeata* subsp. *lucidea* HAMPSON (ex errore pro *leucidia*), Cat. Lepidopt. Phal. Brit. Mus., Suppl. **2**:541.

D i a g n o s i s: Fore coxae alone with orange pilosuty; fore wings in female without cell-spot; discal band much narrower from vein II to dorsum; hind wings with the discal band not narrowing towards costa and bearing six or seven white spots, the last one small, obsolescent and sprinkled with blue.

D i s t r i b u t i o n: New Hanover.

### Damias chalybeata rookensis JORDAN

1915 *Damias chalybeata rookensis* JORDAN, Novit. Zool., **22**:274; fig. 2 (♂). – T e r r a t y p i c a: Rooke Island (Type, ♂, in British Museum).

D i a g n o s i s: Fore wings with a black discoidal spot defined on inner side by a blue white spot or scales conjoined to a patch extending from the lower cell-angle to dorsum, and to a white patch beyond the lower cell angle, between the basas of veins II to V, divided by slight black streaks on the veins. This race is monomorphic.

D i s t r i b u t i o n: Rooke Island.

### Damias chalybeata diserta JORDAN

1926 *Damias chalybeata diserta* JORDAN, Novit Zool., **33**:359. – T e r r a t y p i c a: New Ireland (Types, ♂, ♀, in British Museum).

D i a g n o s i s : Only fore coxae with some orange. Wings of male as in *D. ch. chalybeata* on the upperside, but on the underside of fore wings a continuous bluish white band, mostly widening behind, and in cell a bluish white spot; in female, fore wings with the discal spot nearly even wide from the subcostal to near anal margin, bluer towards the anal margin.

D i s t r i b u t i o n : New Ireland.

## Genus **Cruria** JORDAN

(fig. 15)

1896 *Cruria* JORDAN, Novit. Zool., **3**:39. – S p e c i e s t y p i c a : *Agarista donowani* BOISDUVAL (by original designation).

D i a g n o s i s : Proboscis fully developed; antennae slightly dilated between middle and extremity; palpi upturned; 1st and 2nd joints with long hair in front; 3rd elongate, porrect, naked; frons with a truncate conical prominence, with a raised rim at the extremity; femora with rough hair beneath; tibiae smoothly scaled; abdomen with lateral hair tufts from the middle in male. Venation: in fore

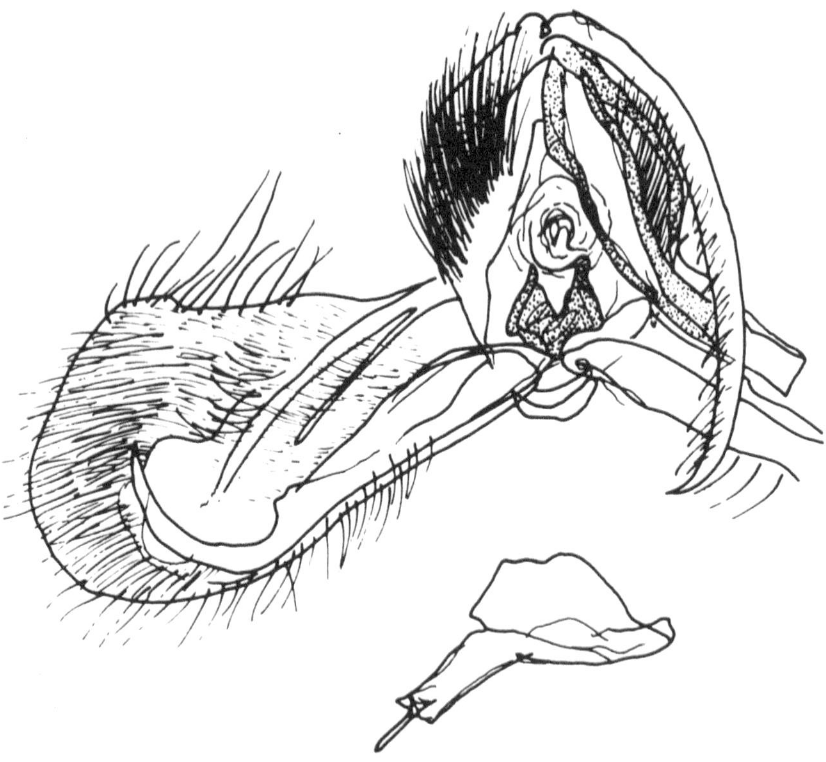

Fig. 15. *Cruria donowani* (BOISDUVAL). Prep. S. G. Kiriakoff N⁰ 14 1970

wings, vein 2 from 3/4 of cell; 3, 4, 5 well separated; distance between 3 and 4 about twice as wide as that between 4 and 5; 6 just below the angle of cell; areole small; 5 and the very short stalk of 10, 8 + 9 from extremity of areole. In hind wing, vein 2 from about 5/7 of cell; 3 and 4 well separated; discocellulars incurved; 5 obsolescent, from middle of discocellulars; 6, 7 from a point.

Male genitalia: Uncus slightly shorter than valva, slender, regularly arched, with a terminal hook. Vincula broadening at middle, bearing a broad bunch of long, stiff hair. Valva rather narrow at base, then broad, lobe-shaped; costa straight or very faintly incurved; apex not marked; termen very broadly rounded; sacculus very faintly incurved for two-thirds, then broadly arched to apex; fold of sacculus rather narrow, reaching to middle of valva to form a rather short, well arched harpe. Aedeagus shorter than fold of sacculus, slender, with the proximal part spoon-shaped and bent at an angle of about 135°; a terminal short, very slender, nearly needle-shaped process.

D i s t r i b u t i o n: Australia; Papua; Timor.

## Key to the species

1 Coxae with grey or yellow pilosity; abdomen banded with pale yellowish . . . . . . . . . 2
1' Mid and hind coxae black; abdomen banded with orange . . . . . . . . . . . . . . . . . . 4
2 Mesothorax without transversal medial line; fore wings without subcellular streak . . . . *C. donowani*, p. 88
2' Fore wings with a subcellular streak in basal area, extending on mesothorax as a transversal line . . . . . . . . . . . . . . . . . . . . . . . . . . . . 3
3 Fore wings with the markings yellowish . . . . . . . . . . . . . . . . . . *C. synopla*, p. 87
3' Fore wings with the markings pure white . . . . . . . . . . . . . . . *C. sthenozona*, p. 91
4 Markings of fore wings white . . . . . . . . . . . . . . . . . . . . . . . *C. neptioides*, p. 89
4' Markings of fore wings yellowish or yellow . . . . . . . . . . . . . . . . . . . . . . . . 5
5 First and second urites with white lines; fore wings without a subbasal fascia in interval I *C. darwiniensis*, p. 90
5' First and second urites without white lines; an orange basal band; fore wings with a diffused whitish subbasal line in interval I . . . . . . . . . . . . . . . . *C. epicharita*, p. 91

## Cruria synopla TURNER

1903 *Cruria synopla* TURNER, Trans. R. Soc. South Australia, Adelaide, 27:1. – T e r r a t y p i c a: Queensland (Types, ♂, ♀, in Australian Museum, Sydney).

D i a g n o s i s: Body and wings blackish brown; palpi with yellowish lateral stripes and some white hair in front; sides of frons, postantennal spots and throat yellowish; collar and metathorax edged with yellow; pectus, coxae and femora with orange pilosity; abdomen with yellowish bands, narrow on basal urites; in female, seventh urite broadly yellow; anal tuft orange at extremity. Fore wings with slight yellow streaks on basal half of costa and in cell; an oblique discocellular bar; two postmedial spots below costa, a spot in interspace II, a point in interval III, and dentate spots nearer to termen in interspace IV and V; three slight greyish subterminal spots below costa; a yellowish spot above tornus; cilia yellowish at the extremity, and with three white dots towards tornus. Hind wings with a deep yellow maculate medial band; cilia yellowish apically, with three white dots below the middle. Length of fore wing 24 – 26 mm.

E a r l y s t a g e s: Larve said to feed on *Calocasia*.

D i s t r i b u t i o n: Australia (Queensland, New South Wales).

### Cruria synopla synopla TURNER

1912 *Cruria synopla synopla* – JORDAN in SEITZ, Großschmetterl. Erde, 11:14.

D i a g n o s i s: Hind wings with the yellow band about half as wide as the terminal black area.

D i s t r i b u t i o n: New South Wales, Southern Queensland (Brisbane, Moreton Bay).

### Crucia synopla latifascia JORDAN

1912 *Cruria synopla latifascia* JORDAN in SEITZ, Großschmetterl. Erde, 11:14 (♂). – T e r r a t y p i c a: Atherton, Southern Queensland (Type, ♂, in British Museum).

D i a g n o s i s: Fore wings with the markings nearly white; hind wings with the yellow band about as broad as the black terminal area.

D i s t r i b u t i o n: Southern Queensland (Atherton); IV or V.

### **Cruria donowani** (BOISDUVAL)

1832 *Agarista donowani* BOISDUVAL, Voyage Astrolabe, Lepidopt.:176. – T e r r a t y p i c a: Mackay, Queensland (Type, ♂, in British Museum).
1892 *Phalaenoides donowani* – KIRBY, Syn. Cat. Lepidopt. Het.:23.
1896 *Cruria donowani* – JORDAN, Novit. Zool., 3:39.
1901 *Cruria donovani* HAMPSON (ex errore pro *donowani*), Cat. Lepidopt. Phal. Brit. Mus., 3:546; fig. 241 (♂).

D i a g n o s i s: Body and wings black; palpi streaked with pale yellow laterally; sides of frons and throat yellow; occiput marked with yellow; edge of collar, a band across mesothorax and tegulae yellow; pectus, coxae and femora with orange pilosity; tarsi ringed with black and white; abdomen with yellow or orange yellow bands; anal tuft orange at the extremity. Fore wings with an oblique white streak below middlle of cell to base of dorsum, sometimes also a streak above middle of vein 1; a white band across cell-end; two more or less obsolete spots beyond the lower cell-angle in interspaces II and III; postmedial spots in interspace V and VI, and IV and V, the latter two more distal; an obsolescent spot near tornus, sometimes absent; three subterminal slightly violaceus spots below costa; cilia white apically, and spotted with white towards tornus. Hind wings with an oblique maculate yellow band, from below costa beyond middle, to mid-dorsum; spot in interspace VII more distal; cilia white apically, and with white points towards tornus. Length of fore wing 23 – 26 mm.

D i s t r i b u t i o n: North West Australia, Queensland, Timor.

E a r l y s t a g e s: Larva brown; each segment ringed with white; a white band bearing several black dots, each dot with a long hair. On *Vitis antarctica*.

### Cruria donowani donowani (BOISDUVAL)

1912 *Cruria donowani donowani* – JORDAN in SEITZ, Großschmetterl. Erde, 11:14; pl. 2f.

Diagnosis: Abdominal rings I–VI white yellowish; hind coxae greyish yellow; postmedial maculate band reaching the base of interspace III.

Distribution: Queensland (Mackay = Terra typica).

### Cruria donowani xanthosoma JORDAN

1912 *Cruria donowani xanthosoma* JORDAN in SEITZ, Großschmetterl. Erde, 11:14 (♂, ♀). – Terra typica: Northwestern Australia (Types, ♂, ♀, in British Museum).

Diagnosis: Abdominal rings, except on urites I and II, orange yellow; abdomen on underside only partly blackish at base in both sexes.

Distribution: North West Australia, western of the Gulf of Carpentaria.

### Cruria donowani timorica JORDAN

1912 *Cruria donowani timorica* JORDAN in SEITZ, Großschmetterl. Erde, 11:14 (♂, ♀). – Terra typica: Delhi, Portuguese Timor (Types, ♂, ♀, in British Museum).

Diagnosis: Abdominal rings yellowish white above, ring on urite VI below as orange as that on urite VII; in female, abdomen on underside black banded with orange; hind wings with the postmedial maculate band narrower, more distal, broken on first radial, especially in female.

Distribution: Portuguese Timor (Delhi).

### Cruria neptioides (BUTLER)

1875 *Agarista neptioides* BUTLER, Ann. Mag. Nat. Hist., London, (4) 15:138 (♀). – Terra typica: Port Albany, North Australia (Type, ♀, in British Museum).
1892 *Phalaenoides neptioides* – KIRBY, Syn. Cat. Lepidopt. Het.:23.
1896 *Cruria neptioides* – JORDAN, Novit. Zool., 11:39.

Diagnosis: Body and wings black; antennae with white points; sides of palpi and of frons, spots on vertex and throat white; collar edged with ochreous in male; fore coxae and femora with orange pilosity; tarsi dotted with white; abdomen banded with orange, except on urites I and II; anal tuft orange, except at base. Fore wings with an orange costal streak in male; some white sprinkling below base of cell; a spot in middle of cell, white like the rest of the markings; a bar across the extremity of cell; a postmedial series of elongate spots, those in interspaces IV and V more distal; two or three subterminal bluish subcostal spots; cilia white apically and with white spots towards tornus. Hind wings with an oblique band of white spots; veins black on it; cilia white apically and with series of white spots. Length of fore wing 22 – 25 mm.

Distribution: Queensland, Papua.

### Cruria neptioides neptioides (BUTLER)

1912 *Cruria neptioides neptioides* – JORDAN in SEITZ, Großschmetterl. Erde, **11**:14; p. 2f.

Diagnosis: Hind wings with a white cell-spot; band as broad as the terminal black band; in female, band shorter.

Distribution: North Australia (Cape York, Thursday Island).

### Cruria neptioides papuana JORDAN

1912 *Cruria neptioides papuana* JORDAN in SEITZ, Großschmetterl. Erde, **11**:14 (♂). – Terra typica: Aroa River, Papua (Type, ♂, in British Museum).

Diagnosis: Hind wings without a white cell-spot; white band slightly narrower; in female, white markings much reduced, especially the band of hind wings.

Distribution: Papua (Aroa River, Port Moresby); XI. till III.

### Cruria darwiniensis (BUTLER)

1884 *Agarista darwiniensis* BUTLER, Ann. Mag. Nat. Hist., London, (5) **14**:406 (♀). – Terra typica: Port Darwin, North Australia (Type, ♀, in British Museum).
1892 *Phalaenoides darwiniensis* – KIRBY, Syn. Cat. Lepidopt. Het.:23.
1896 *Cruria darwiniensis* – JORDAN, Novit. Zool., **3**:39.
1901 *Cruria darviniensis* HAMPSON (ex errore pro *darwiniensis*), Cat. Lepidopt. Phal. Brit. Mus. **3**:547; pl. 52, fig. 14.

Diagnosis: Body and wings black; sides of palpi and of frons white; occiput spotted with yellow; egdes of collar and throat yellow; fore coxae and femora with orange pilosity; tibiae streaked with white; tarsi ringed with white and black; first and second urites with white lines, rest banded with orange; extremity orange. Fore wings with the markings pale yellow, sometimes ochreous: a basal point; a subcostal streak on basal half; a spot in cell-middle; an oblique bar across cell-extremity; postmedial spots in interspaces V and VI, and more distally in interspaces IV and V; tornus with an obsolete spot, two or three bluish white subterminal spots below costa; cilia white apically and spotted with white towards tornus. Hind wings with an oblique pale yellow or ochreous band of spots from below costa beyond middle to middle of anal margin; sometimes a distinct, sometimes an obsolete white cell-spot; cilia white apically, and with series of white spots. Length of fore wing 21 – 27 mm.

Distribution: North West Australia and Queensland.

### Cruria darwiniensis darwiniensis (BUTLER)

1912 *Cruria darwiniensis darwiensis* – JORDAN in SEITZ, Großschmetterl. Erde, **11**:14; pl. 2f.

D i a g n o s i s: Markings pale yellow; hind wings in male with only traces of cell-spots.

D i s t r i b u t i o n: Port Darwin (Terra typica); West of Gulf of Carpenteria.

## Cruria darwiniensis platyxantha (MEYRICK)

1891 *Agarista platyxantha* MEYRICK, Trans. R. Soc. S. Australia, Adelaide, **14**:194. – T e r r a t y p i c a: Queensland (Type in British Museum).
1891 *Agarista tropica* LUCAS, Proc. Linn. Soc. New South Wales, Sydney, (2) **6**:302. – T e r r a t y p i c a: Queensland (Type in South Australian Museum, Adelaide).

D i a g n o s i s: Wings with the markings ochreous; hind wings with a bar-shaped ochreous spot in cell, generally absent in female.

D i s t r i b u t i o n: Queensland, South of Cooktown (Cedar Bay).

## Cruria epicharita TURNER

1911 *Cruria epicharita* TURNER, Ann. Queensland Mus. Brisbane, **10**:159 (♂). – T e r r a t y p i c a: Cape York, Queensland (Type, ♂, in Australian Museum, Sydney).

D i a g n o s i s: Body and wings blackish brown; patch between antennae, sides of frons and throat orange; palpi with some orange hair in front at base; collar with orange bars; fore coxae and femora with some orange pilosity; abdomen with an orange bar at base and six narrow bands; anal tuft orange at the tip. Fore wings with a diffused whitish subbasal streak in interspace I; a yellowish white patch at cell-extremity; elongate marks above veins 2, 3, 6, 7 in submedian region, and marks more distally in interspaces IV and V; slight subterminal marks in interspaces III, VI and VII; cilia white apically. Hind wings with a very broad yellowish white band from terminal portion of cell and in interspace V to the anal margin, rather irregular on its edges; cilia white apically. Length of fore wing 24 mm.

D i s t r i b u t i o n: Queensland (Cape York).

## Cruria sthenozona TURNER

1920 *Cruria sthenozona* TURNER, Trans. R. Soc. South Australia, Adelaide, **44**:124. – T e r r a t y p i c a: North Queensland (Type in Australian Museum, Sydney).

D i a g n o s i s: Resembles *Cruria synopla*, but the marks of fore wings without ochreous tinge; a slender line across the thorax behind the middle, extended into fore wings beneath cell to about middle. Hind wings with a very broad ochreous-whitish median band, nearly half the breadth of the wing, not dislocated at the costal extremity towards the tornus. Length of fore wing 14 – 15 mm.

D i s t r i b u t i o n: North Queensland (Terra typica = Atherton, Evelyn scrub near Herberton); V, X and I.

## Genus **Cruriopsis** JORDAN

(fig. 16)

1912 *Cruriopsis* JORDAN in SEITZ, Großschmetterl. Erde, **11**:21–22. – Species typica: *Eusemia funebris* MOORE (by original designation).

Diagnosis: Proboscis well developed. Antennae filiform, progressively slenderer and compressed distally, and with setae in both sexes; these longer than the antenna is broad. Palpi upturned, with the second joint fringed with hair, and the last joint short in both sexes. Eyes naked. Frons with a truncate conical prominence, without raised rim at the extremity. Tibiae with appressed scales. Venation: in fore wings, vein 2 from about five sixths of cell; 3, 4, 5 separated, nearly equidistant; discocellulars oblique, incurved; 6 from angle of cell; areole very narrow; 7 and the moderately long stalk of 8 + 9 from the extremity of areole; 10 from upper angle of areole. In hind wings, vein 2 from three-quarters of cell; 3, 4

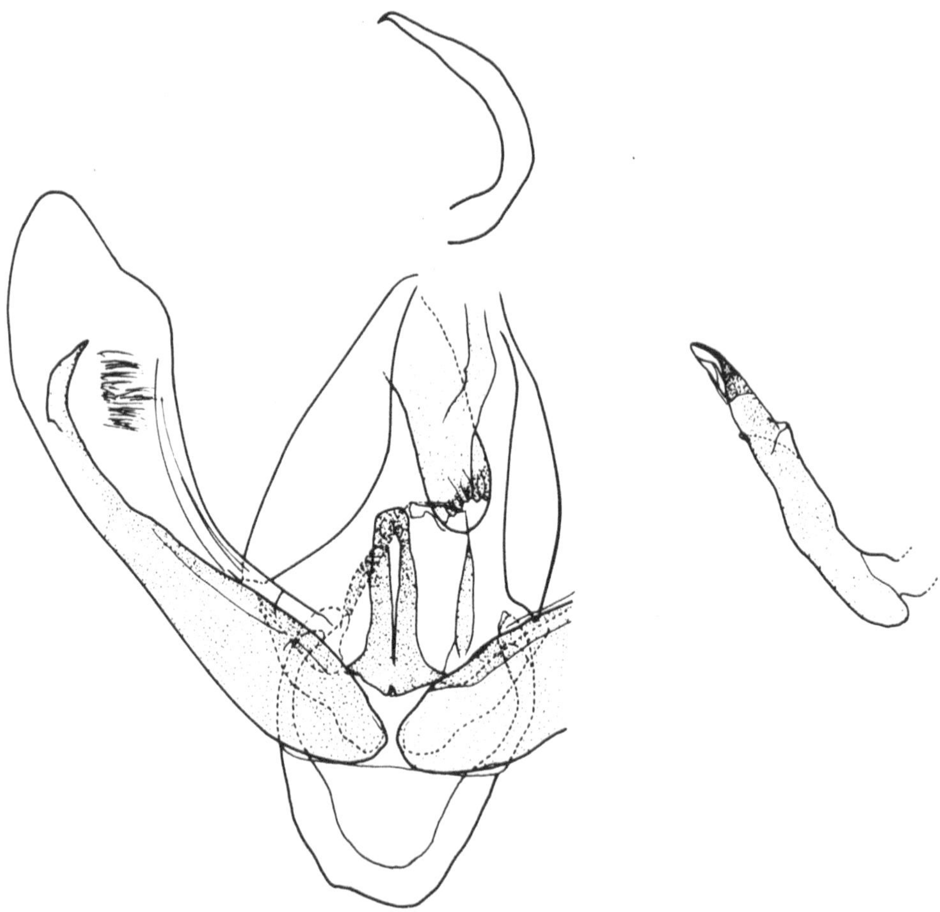

Fig. 16. *Cruriopsis funebris* (MOORE). Type. B. M. Noct. 8451

and 6, 7 from the cell-angles; discocellulars incurved in the lower portion, nearly straight in the upper one; 5 obsolescent, from middle of discocellulars.

Male genitalia: Uncus narrow, about two-thirds of costa in length, well arched, ending in a minute hook. Tubus analis short, moderately broad, sclerified. Folds of vinculum broad in the middle portion. Valva elongate, rather narrow, but broadening in distal quarter; costa incurved for three-quarters of its lenght, then slightly prominent; apex rounded; sacculus outcurved, more distinctly at base and subterminally; fold of sacculus very broad, almost reaching the costal margin, narrowing from middle of valva, to end in a very short arched harpe placed at three-quarters of valva. Aedeagus rather short, about half of valva in length, nearly straight, tapering to a point distally; fultura inferior with the proximal angles produced, otherwise narrow, and reaching nearly to middle of vinculum. Saccus broadly rounded.

D i s t r i b u t i o n: China, Indo-China, Northern India.

## **Cruriopsis funebris** (MOORE)

1872 *Eusemia funebris* MOORE, Proc. Zool. Soc. London, **1872**:569 (♂, ♀). – T e r r a t y p i c a: Darjeeling, Sikkim (Type, ♂, in Berlin Museum).
1892 *Phalaenoides funebris* – KIRBY, Syn. Cat. Lepidopt. Het.:24.
1901 *Ophthalmis funebris* – HAMPSON, Cat. Lepidopt. Phal. Brit. Mus., 3:649–650; fig. 285 (♂, ♀).
1912 *Cruriopsis funebris* – JORDAN in SEITZ, Großschmetterl. Erde, **11**:22; pl. 3f.

D i a g n o s i s: Head, body and wings black; palpi and sides of frons spotted with white; pectus with orange pilosity; tarsi ringed with white; abdomen with a white band at base and segmental orange bands; anal tuft mostly orange; underside orange, spotted with black subterminally. Fore wings with a yellowish dot in middle of cell; an ochreous-white bar near cell-end; a spot beyond the upper cell-angle, and two spots beyond the lower cell-angle; blue-green dots at base, subbasally, on disk and subterminally; cilia spotted with white in apical area. Hind wings with a broad oblique white band in proximal half, reduced in female; a white spot beyond the upper cell-angle; and a series of white spots between the anal margin and vein 4; cilia spotted with white. Length of fore wing 26 – 35 mm.

D i s t r i b u t i o n: Assam, Sikkim, Indochina, South, West and East Central China.

### C r u r i o p s i s  f u n e b r i s  f u n e b r i s (MOORE)

1912 *Cruriopsis funebris funebris* – JORDAN in SEITZ, Großschmetterl. Erde, **11**:22; pl. 3f.

D i a g n o s i s: As above.

D i s t r i b u t i o n: Sikkim, Assam.

## Cruriopsis funebris vithoroides (LEECH)

1890 *Eusemia vithoroides* LEECH, Entomologist, London, **16**:8; pl. 1, fig. 4 (♀). – Terra typica: Changyang, China (Type, ♂, in British Museum).
1892 *Episteme vithoroides* – KIRBY, Syn. Cat. Lepidopt. Het.:29.
1901 *Ophthalmis funebris* subsp. *vithoroides* – HAMPSON, Cat. Lepidopt. Phal. Brit. Mus., 3:650.
1912 *Cruriopsis funebris vithoroides* – JORDAN in SEITZ, Großschmetterl. Erde, **11**:22.

Diagnosis: Fore wings with the white markings larger, cell-spot triangular; bluish markings more developed; hind wings with the white discal band divided by a black line; larger: length of fore wing about 29 mm in male, and 35 mm in female.

Distribution: China (Szechuan, Southern Hupé: Changyang = terra typica; Chekiang.

## Cruriopsis funebris cognata JORDAN

1912 *Cruriopsis funebris cognata* JORDAN in SEITZ, Großschmetterl. Erde, **11**:22 (♂). – Terra typica: Horisha, Taiwan (Type, ♂, in British Museum).

Diagnosis: Fore wings with the spots beyond the lower cell-angle larger and white; hind wings with the white markings better developed than in *C. f. vithoroides*; the discal band broader than the black medial band.

Distribution: Southeastern China (Taiwan; Horisha = terra typica).

## Genus **Comocrus** JORDAN

(fig. 17)

1896 *Comocrus* JORDAN, Novit. Zool. 3:40. – Species typica: *Agarista behri* ANGAS (by original designation).

Diagnosis: Proboscis fully developed; antennae distinctly dilated between middle and extremity; palpi obliquely upturned; 1st and 2nd joints with long pilosity in front; 3rd joint long, naked, porrect; frons with a conical projection bearing at its truncate extremity a raised rim; eyes naked; mid and hind tibiae with long hair at the outer edge; tarsi smooth; abdomen clothed with long hair. Venation: in fore wings, vein 2 from 5/7 of cell; 3, 4 and 5 separated, about equidistant; 6 from just below the cell-angle; areole small, narrow; 7 and stalk of 10, 8 + 9 from extremity of areole. In hind wings, vein 2 from slightly beyond 2/3 of cell; 3, 4 and 6, 7 from cell-angles; 5 obsolescent, from the middle of discocellulars.

Male genitalia: Uncus rather short, slender, weakly arched, pointed. Tubus analis shorter than uncus, spoon-shaped. Vincula broad, with at each side two bunches of long, stiff hair, one medio-distal and one latero-proximal. Valva banana-shaped, but not over-elongate; costa slightly less arched than sacculus; apex broadly rounded; basal fold of sacculus broad, but very short, reaching only at about one-quarter of costa, and ending in a short, slender, arched harpe; a row of minute stiff setae from

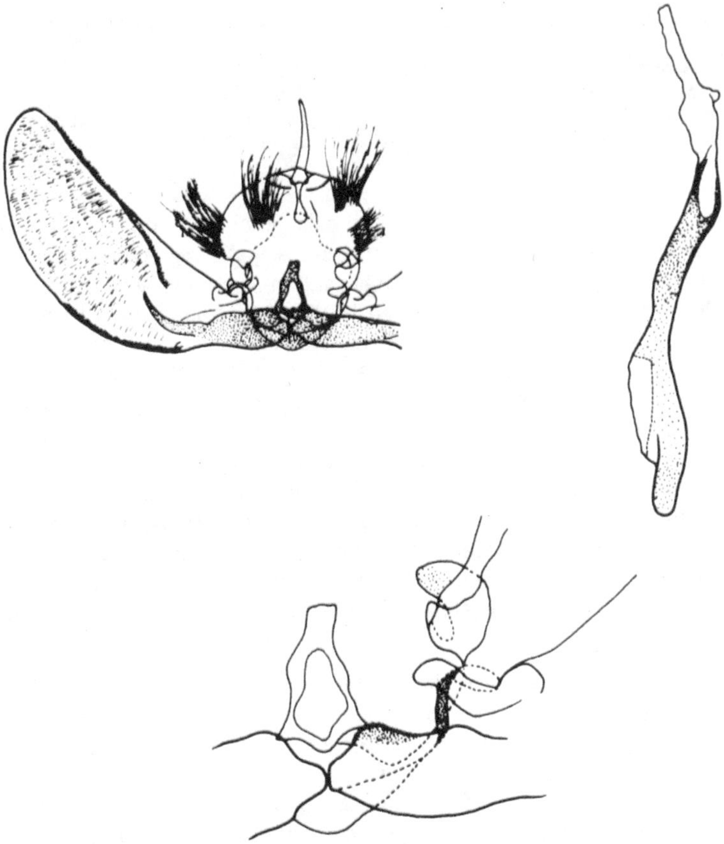

Fig. 17. *Comocrus behri* (ANGAS). BM Noct. 8388

near base in costal part of valvula, running upwards, reaching costa at middle, and ending at apex; a similar row along sacculus, from end of fold to two-fifths. Aedeagus one half again as long as costa, robust proximally, then slender, very faintly arched to form a S; terminally, a finger-shaped process; fultura inferior conical-truncate. Saccus hardly differentiated.

D i s t r i b u t i o n : Australia.

## Comocrus behri (ANGAS)

1847 *Agarista behri* ANGAS, South Austral. Illustr.: pl. 37, fig. 21, 22. – T e r r a t y p i c a : Australia (Type not designated).

1864 *Agarista contorta* WALKER, List Lepidopt. Het. Brit. Mus., **31**:45 (♀). – T e r r a t y p i c a : Sydney, New South Wales (Type, ♀, in British Museum).

1864 *Agarista flexuosa* WALKER, List Lepidopt. Het. Brit. Mus., **31**:46 (♀). – T e r r a t y p i c a : Moreton Bay, Queensland (Type, ♀, in British Museum).

1865 *Agarista casuarina* SCOTT, Austral. Lepidopt., 1:24; pl. 8. – Terra typica: Australia.
1892 *Phalaenoides contortus* – KIRBY, Syn. Cat. Lepidopt. Het.:22.
1892 *Phalaenoides flexuosus* – KIRBY, Syn. Cat. Lepidopt. Het.:22.
1896 *Comocrus cortortus* JORDAN (ex errore pro *contortus*), Novit. Zool., 3:40.
1901 *Comocrus behri* – HAMPSON, Cat. Lepidopt. Phal. Brit. Mus., 3:544–545; fig. 240 (♂, ♀).

Diagnosis: Black; body marked with pale yellow, viz., on palpi, sides of frons, vertex, collar, tegulae, middle of thorax above, gular region and forepart of pectus; pilosity of fore legs rufous buff; tibiae and tarsi spotted with white; ventral bands on the two subterminal urites and anal tuft rufous orange. Fore wings with pale yellow markings, viz.: an oblique subbasal line reaching vein 1; an oblique, nearly straight inner line; a much broader band from middle of costa to lower angle of cell; connected to it, a narrow band, not reaching the outer line; a narrow band from three fifths of costa to vein 4; outer line very distal, from about three-quarters of costa to near tornus, forming two strong teeth on vein 4 and in interval I; veins streaked with white in terminal area; cilia white, spotted with black except on apex and tornus. Hind wings black with faint blue reflections; cell slightly marked with yellow; a discocellular spot, sometimes absent; a postmedian series of spots, angled outwards on vein 4, sometimes forming a nearly continuous band, sometimes absent; cilia white, spotted with black, except in apical area. Length of fore wing 26 – 35 mm.

Early stages: **Larva** black; head, first thoracic tergite, legs and prolegs orange; each segment from the second to the penultimate, with two white bands, white markings and white spiracles; penultimate ring red dorsally and laterally; last ring with a white band; each segment bearing several white setae dilated and flattened at tip. When at rest, head and thorax raised and bent backwards; when disturbed, larva emits from mouth a yellowish green fluid. On *Loranthus pendulus* and *Casuarina*. – **Pupa** in a strong cocon of pieces of bark, dry leaves and earth.

Distribution: Australia (New South Wales, Queensland, West Australia).

## Genus **Crinocula** JORDAN

(fig. 18)

1896 *Crinocula* JORDAN, Novit. Zool., 3:36; pl. 40. – Species typica: *Crinocula kinabaluensis* ROTHSCHILD (by original designation).

Diagnosis: Proboscis strongly developed; antennae distinctly dilated before the extremity; palpi upturned; 1st and 2nd joints with long hair in front, 3rd short, smooth, hardly twice as long as broad, porrect, frons narrow, without prominence; eyes with a long and dense pubescence; body clothed with rough hair; tibiae and tarsi with appressed scales; abdomen of male with lateral tufts of long hair from the second urite. Nervation: Fore wings with vein 2 from about 7/10 of cell; 3, 4, 5 approximated, about equidistant; discocellulars incurved; 6 from upper angle; areole very small, sometimes absent in female; 7, 10 and 8 + 9 from extremity of areole. Hind wings with vein 2 from 3/4 of cell; 3, 4 and 6, 7 from angles of cell; discocellulars incurved; 5 obsolescent from middle of discocellulars.

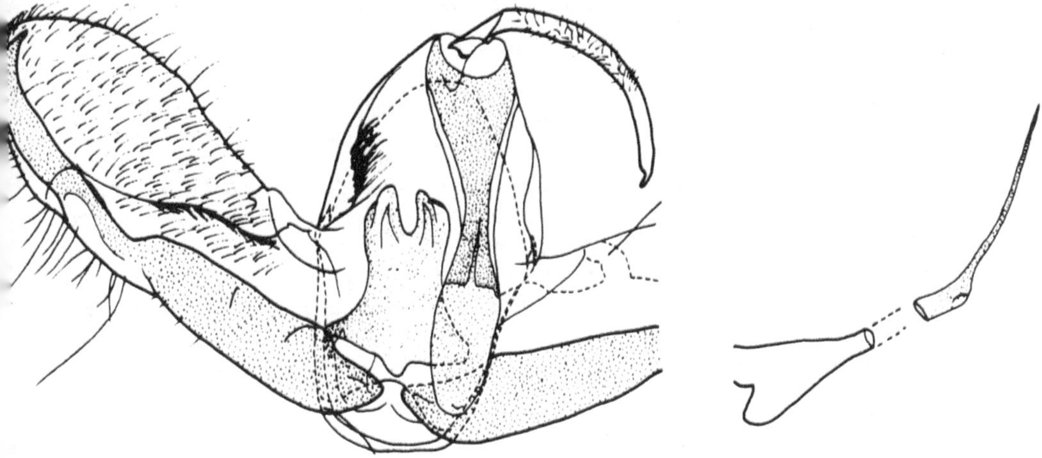

Fig. 18. *Crinocula kinabaluensis* ROTHSCHILD. Type. B. M. Noct. 8377

Male genitalia: Uncus moderately long, about half of sacculus, slender, regularly arched, with short pilosity and a small terminal hook. Vinculum broad, fringed in proximal half with rather short stiff hair. Valva elongate, rather broad, lobe-shaped; costa faintly arched; apex rounded; fold of sacculus moderately broad, reaching to about middle of valva and ending in a long, robust harpe, arched terminally and reaching the apex of valva. Aedeagus short, about one-third of sacculus, robust proximally, but soon becoming slender, nearly straight, and ending in a needle-shaped, faintly arched process, nearly as long as aedeagus; fultura inferior strongly developed, broad proximally, with the sides incurved, and ending in two broad lobes. Saccus very short, hardly differentiated, rounded.

Distribution: Borneo.

## **Crinocula kinabaluensis** ROTHSCHILD

1896 *Crinocula kinabaluensis* ROTHSCHILD, Novit. Zool., 3:36; pl. 14, fig. 22 (♂, ♀). – Terra typica: Kina Baloe, Borneo (Types, ♂, ♀, in British Museum).
1901 *Crinocula cinabaluensis* HAMPSON (ex errore pro *kinabaluensis*), Cat. Lepidopt. Phal. Brit. Mus., 3:530; fig. 232.

Diagnosis: Head and antennae black, mixed with yellow hair, especially on sides of palpi and frons; thorax above black; thorax below and legs orange; legs mixed with black; abdomen orange with a black dorsal stripe and extremity; in female only a basal black spot is present. Fore wings black, with a few blue scales; an oblique orange band just beyond the cell, running from near costa to vein 2. Hind wings black at base and in distal half; a broad median band of red. Length of fore wing 17 – 22 mm.

Distribution: North Borneo (Kina-Baloe = terra typica).

## Genus **Saigonita** KIRIAKOFF

(fig. 19)

1971 *Saigonita* KIRIAKOFF, Lambillionea, **56**:5–6. – Species typica: *Saigonita paradoxa* KIRIAKOFF (by original designation).

Diagnosis: Proboscis fully developed; antennae dilated towards extremity; palpi short, obliquely upturned; 1st and 2nd joints fringed with long hair; 3rd short, hidden; femora and tibiae clothed with long hair; hind tibiae with two pairs of spurs, the proximal pair the longest. Fore wings with the apex produced and rounded; costa slightly concave before middle, then outcurved; termen oblique, mainly straight, but slightly incurved below the apex; venation: vein 2 from four-fifths of cell; 3, 4, 5 much approximated, equidistant; 6 from upper angle of cell; areole absent; 7 short

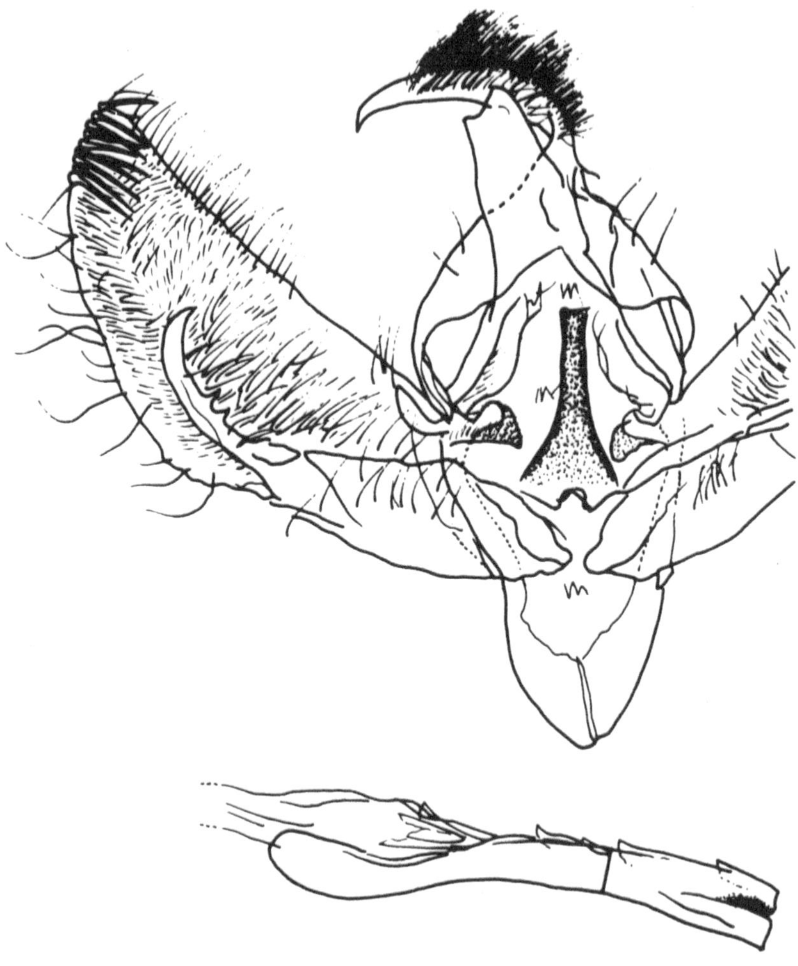

Fig. 19. *Saigonita paradoxa* KIRIAKOFF. Type Paris Mus. prep. P. Viette

stalked with the very short stalk of 8 + 9; 10 from upper angle of cell; media dilated, finely ribbed, no doubt used for stridulation. Hind wings with vein 2 very near angle of cell; 3, 4 and 6, 7 from angles of cell; 5 slightly weaker than the other veins, from middle of cell; 8 anastomosing with cell for slightly less than the basal third thereof.

Male genitalia: Uncus elongate, narrow, arched, with a small terminal hook. Tubus analis short, with basal lateral sclerification. Tegumen broadened distally, without stiff hair. Valva elongate, banana-shaped; costa nearly straight; apex upturned, rounded; sacculus well outcurved, especially at middle; fold of sacculus rather broad, ending at middle of valva in a short, arched harpe with a few denticulations basally at distal margin; apex bearing a row of stiff, spinulate hairs. Edeagus as long as the costa, slender, faintly forming an S in distal portion; fultura inferior not sclerified; fultura superior narrowing distally. Saccus semi-elliptical.

Distribution: Vietnam.

## Saigonita paradoxa KIRIAKOFF

1971 *Saigonita paradoxa* KIRIAKOFF, Lambillionea, **71**:7–8; fig. 1 (♂). – Terra typica: Saigon, South Vietnam (Type, ♂, in Paris Museum).

Diagnosis: Head, body and appendages brown. Wings brown, slightly tinged with rufous. Fore wings with the inner and outer lines dark, double, filled in and edged on both sides with orange; inner line running from one fifth of costa to two thirds of dorsum, more or less straight; outer line running from three quarters of costa to tornus; discocellular mark subquadrate, darker than the ground colour; outer line followed by a series of triangular internervural spots that do not reach the termen; cilia brown with a pale median line. Hind wings with a parallelogram-shaped very drak rufous-brown discal spot, starting at costa and reaching the anals, limited proximally and distally by an orange spot; the distal spot nearly cut in two by a projection of the discal spot; basal area greyish brown, merging into orange of the proximal edge of the inner line; distal and anal areas rufous brown; a double postdiscal line, filled in with orange, limiting the orange area, straight but faintly elbowed on the last anal; a terminal series of brown sagittate spots; termen bearing in the anal area a light grey spot, sprinkled with black, and limited proximally and distally by silvery scales; cilia longer in this part of termen and silvery at base, otherwise grey, with black base and extremity; rest of cilia as in fore wings. Length of fore wing 19 mm.

Distribution: South Vietnam (Saigon = terra typica).

## Genus **Ipanica** HAMPSON

(fig. 20)

1896 *Ipana* JORDAN, Novit. Zool., **3**:54 (nec WALKER 1858). – Species typica: *Aegocera cornigera* BUTLER (by original designation).
1910 *Ipanica* HAMPSON, Cat. Lepidopt. Phal. Brit. Mus., **9**:450 (nomen novum).

Diagnosis: Proboscis fully developed; antennae minutely serrate and ciliated; palpi porrect, with the second joint moderately fringed with hair and the third moderate; frons with long flattened corneous process with acute angles, and a square corneous plate below it; thorax with hair only, without crests; tibiae moderately fringed with hair; abdomen without crests. Venation: in fore wings, vein 2 from two-thirds of cell; 3, 4, 5 approximated, equidistant; discocellulars faintly incurved; 6 from upper angle of cell or from the extreme base of areole; the latter rather long, narrow; 7 and the short stalk of 8 + 9 from the extremity of areole; 10 from the upper angle of areole. Hind wings with vein 2 from two-thirds of cell; 3, 4 and 6, 7 from the cell-angles; discocellulars well incurved; 5 obsolescent, from middle of discocellulars.

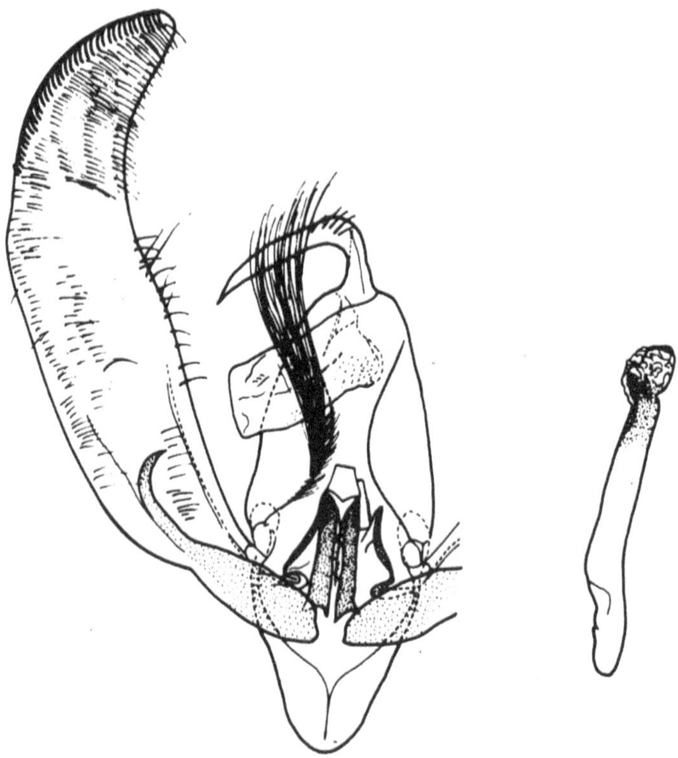

Fig. 20. *Ipanica cornigera* (BUTLER). Type. B. M. Noct. 8395

Male genitalia: Uncus short, less than one-third of costa, bent at a right angle at one-quarter, then nearly straight, tapering to a slightly curved point. Tubus analis short and broad. Fold of vinculum broadly rounded, fringed with short hair and bearing at middle a bunch of very long hair. Valva very elongate, banana-shaped, with the costa very faintly outcurved at middle; apex upturned, rounded; fold of sacculus only reaching to the end of the narrow basal portion of valva, broad, ending in a short sickle-shaped harpe, as long as the straight portion of uncus; fold and harpe forming together less than one-third of valva; terminal fifth of sacculus fringed with short spicule-like hair. Aedeagus about one-third of valva in length, slender, nearly straight, slightly swollen, and bearing laterally a area of minute cornuti in the terminal portion; fultura inferior consisting of two narrow pieces bearing terminally a number of minute warts. Saccus semi-elliptical.

Distribution: Indonesia, New Guinea, Australia.

### **Ipanica cornigera** (BUTLER)

1886 *Aegocera cornigera* BUTLER, Trans Entomol. Soc. London, **1886**:381 (♂). – Terra typica: Gayndah, Queensland (Type, ♂, in British Museum).
1896 *Ipana cornigera* – JORDAN, Novit. Zool., 3:54.
1910 *Ipanica cornigera* – HAMPSON, Cat. Lepidopt. Phal. Brit. Mus., 9:450; fig. 216.
1954 *Ipanica cornifera* KIRIAKOFF (lapsus calami), Verh. K. vlaam. Acad. Wet., **47**:27; fig. 13 (tympanal organs).

Diagnosis: Head and thorax blackish brown; palpi white at base; 2nd joint white at the extremity; head, collar, tegulae and metathorax marked with white; pectus orange; legs orange and brown; extremities of tibiae and the tarsi brown banded with white; abdomen orange, with the edges of the urites brownish, broader on the underside. Fore wings Vandyck-brown; base spotted with white; inner line from one fifth of costa, oblique to one-third of dorsum, forked dorsally; beyond it, a few white spots; cell-extremity with a squarish spot; an oblique line from below cell to near tornus; outer line broad, from two-thirds of costa to tornus; apical area spotted with white; a terminal series of small spots; cilia brown, spotted with white. Hind wings orange; a small blackish discocellular spot; broad black terminal band, strongly excurved and irregular after vein 5; cilia spotted with black and orange. Size variable, length of fore wing 14 mm on an average.

Early stages: Larva yellow or orange, banded with blackish or brown, the median band broken laterally and forming dorsally a U-shaped mark; above the legs a pale streak; head, first thoracal ring above, underside and a transverse spot on the eleventh ring, red.

Distribution: Tenimber Islands, Soemba, New Guinea, North and West Australia, Queensland.

### Genus **Hecatesia** BOISDUVAL

(fig. 21)

1829 *Hecatesia* BOISDUVAL, Monogr. Zygaen.:11. – Species typica: *Hecatesia fenestrata* BOISDUVAL (by monotypy).

Diagnosis: Proboscis fully developed; antennae distinctly to strongly dilated towards the extremity, with the apex obtuse; palpi upturned, with the second joint fringed with long hair in front, and the third somewhat porrect; frons with truncate conical prominence with the extremity raised at the rim, and with a flattened plate below it; tibiae fringed with rough hair above; abdomen with dorsal crests and lateral tufts; male with protrusible lateral tufts of hair from the base, and bearing a stridulating organ on fore wings. Venation: in fore wings, vein 2 from three-quarters of cell; 3, 4 separate; 5 above angle of cell; areole long and very narrow; 6 from upper cell-angle; upper part of discocellulars, between veins 5 and 6 more or less obliterated; in male, veins 7 and 8 from the extremity of areole, 9 and 10 stalked; in female, veins 8 and 9 stalked, and vein 10 from areole. Hind wings with vein 2 from slightly beyond two-thirds of cell; 3, 4 and 6, 7 from the cell-angles; discocellulars incurved; 5 obsolescent, from middle of discocellulars. Male stridulating organ: Apex of forewings oblique truncate-rounded; between the cell and the costalis a hyaline, transversely ribbed, elongate-fabiform spot; costa and submedia bent accordingly; middle of spot thickened on costal margin. The stridulating sound is being produced by rubbing the mid tarsi against the hyaline spot.

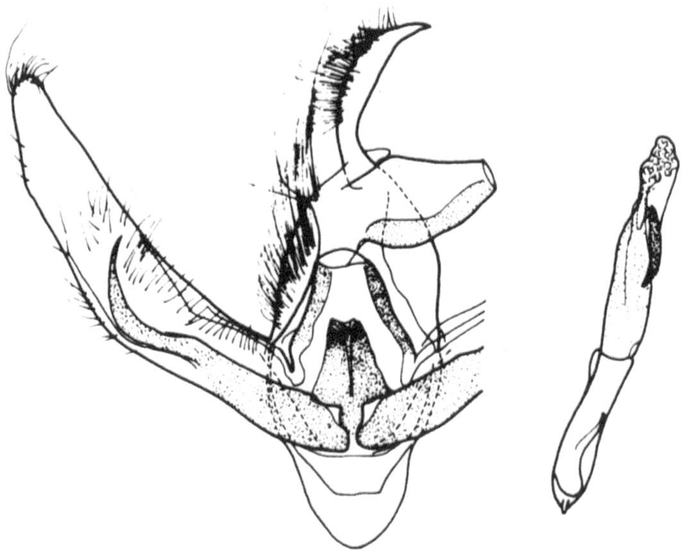

Fig. 21. *Hecatesia fenestrata* BOISDUVAL. B. M. Noct. 8391

Male genitalia: Uncus moderately long, about half the costa, arched, compressed, with the extremity acute. Tubus analis short, sclerified ventrally. Vincula moderate; fold moderately broad, rounded, with long pilosity. Valva relatively very narrow, with costa and sacculus more or less parallel, except terminally where the valva narrows; apex rounded; sacculus very faintly angled beyond middle; fold rather narrow, reaching to about one-third of sacculus and ending in a curved harpe about as long as the uncus. Aedeagus very slightly shorter than costa, rather slender, nearly straight, bearing subterminally a slender, arched process, appressed to the side of the aedeagus and covered with minute cornuti and denticulations; fultura inferior narrowing distally. Saccus semi-elliptical.

D i s t r i b u t i o n : Australia.

## Key to the species

1 Fore wings with numerous small white and yellowish spots ....... ***H. exultans*, p. 103**
1′ Fore wings with two or three maculate bands .................. 2
2 Fore wings with two maculate bands ..................... ***H. fenestrata*, p. 103**
2′ Fore wings with three maculate bands .................. ***H. thyridion*, p. 104**

### Hecatesia fenestrata BOISDUVAL

1829 *Hecatesia fenestrata* BOISDUVAL, Monogr. Zygaen.:11; pl. 1, fig. 1 – 2. – Terra typica: Australia (Type, ♂, in British Museum).

Diagnosis: Head and thorax dark chocolate brown; palpi in front, sides of frons, streaks on vertex and collar and patches on tegulae yellowish white; pectus with brown pilosity; legs blackish, banded with white; abdomen orange with dorsal crests and lateral spots black; some white at base of dorsum; underside brown at base, white at the extremity; lateral tufts and anal tuft black. Fore wings deep chocolate brown; an oblique medial band of yellowish confluent spots, a postmedial maculate band running from the extremity of the hyaline patch to tornus, interrupted on vein 2; a terminal series of white dots. Hind wings orange; costal area broadly dark brown, confluent with the smallish discocellular spot; terminal area dark brown, broader below vein 2; cilia white, spotted with brown. Length of fore wing 12 mm.

Distribution: Victoria to New South Wales.

### Hecatesia exultans WALKER

1864 *Hecatesia exultans* WALKER, List Lepidopt. Het. Brit. Mus., **31**:58 (♀). – Terra typica: West Australia (Type, ♀, in British Museum).

Diagnosis: Head and thorax black, the latter with blue gloss; 2nd joint of palpi at base and at the extremity, sides of frons, occiput, a postocular line, collar at middle and laterally and tegulae at base and extremity, with pale yellow spots; antennae with the extremity white; femora and tibiae with tufts of orange hair; tarsi ringed with white; abdomen black; dorsal crests and lateral and anal tufts orange; last four urites with sublateral white spots. Fore wings black; yellow spots at base; antemedial and medial spots in and below cell; spots beyond cell between veins 3–5 and 5–10; subterminal internervural spots on the apical half, and one above tornus; blue-white spots in and below cell before middle, on discocellulars, and postmedial dots towards costa and beyond the lower cell-angle; diffused patches of scales below the cell-end and on dorsal area; cilia spotted black and white. Hind wings orange; a small discocellular black spot; a terminal black band, expanding on apical area, and on costa extending to before middle; cilia spotted black and white. Length of fore wing 13 mm.

Distribution: West Australia.

## Hecatesia thyridion FEISTHAMEL

1839 *Hecatesia thyridion* FEISTHAMEL, Rev. Mag. Zool., (2) 1:2; pl. 21, fig. 1. – Terra typica: King Georg's Sound, West Australia (Type, ♂, in British Museum).

Diagnosis: Head and thorax dark chocolate brown; antennae ringed with white; palpi in front, sides of frons, stripes on occiput, collar and tegulae pale yellow; pectus with brown pilosity; legs blackish brown, ringed with white, and with tufts of orange hair on fore tibiae; abdomen orange; dorsal crests and lateral spots black; some white at the base of dorsum; underside brown at base, banded black and white on terminal half; anal tuft blackish at tip. Fore wings dark chocolate brown, still darker in female; an oblique subbasal yellowish bar from below costa to below cell, and a spot above vein 1; an oblique medial row of four spots; a row of three spots from below costa beyond middle to tornus; the ribbed hyaline spot extending from the medial to the postmedial spots; five white spots on apical part of termen. Hind wings orange; costal area broadly black from near base and confluent with the large discocellular spot; terminal area black, broader apically and below vein 2; a terminal line below vein 2 orange; cilia orange; minute white dots in the interspaces. Length of fore wing 16 mm.

Distribution: West Australia (King George's Sound = terra typica).

## Genus Argyrolepidia HAMPSON

(fig. 22)

1901 *Argyrolepidia* HAMPSON, Cat. Lepidopt. Phal. Brit. Mus., 3:550. – Species typica: *Geometra pamphilia* STOLL (by original designation).
1920 *Odice* HAMPSON, Cat. Lepidopt. Phal. Brit. Mus., Suppl. 2:543 [nec HÜBNER, Verzeichn. exot. Schmetterl., (17) 258, 1823]. – Species typica: *Geometra acharia* CRAMER, designated by JORDAN 1912.

Diagnosis: Proboscis fully developed; antennae setiform, not dilated distally, but somewhat compressed; palpi upturned, with the second joint fringed with hair, the third long, naked, somewhat porrect and dilated at the extremity; eyes naked; frons with a somewhat pointed conical projection, sometimes with a raised ring; pectus hairy; femora fringed with long hair; tibiae nearly smooth; abdomen clothed with ridges of hair, in male with protrusible lateral tufts of hair at the base. Venation: in fore wings, vein 2 from before three-quarters of cell; 3, 4 from the lower cell-angle; 5 distinctly above the angle; upper discocellular straight, oblique; lower portion nearly vertical; 6 from the upper cell-angle; areole narrow; 7 and the long stalk of 8 + 9 from its extremity; 10 from upper margin of areole, near the extremity. In hind wings, vein 2 from two-thirds of cell; 3, 4 and 6, 7 from the cell-angles; discocellulars faintly incurved; 5 obsolescent, from middle of discocellulars.

Male genitalia: Uncus slightly shorter than half of costa, very narrow, nearly filiform, with a minute terminal hook. Tubus analis weakly sclerified on the ventral side. Vinculum with lateral folds, broad proximally, gradually narrowing distally, with a large basal tuft of stiff hair. Valva elongate and narrow; costa nearly straight in the proximal half, then outcurved, and finally faintly incurved subterminally; apex pointed; sacculus outcurved, with a terminal row of stiff spicule-like hair;

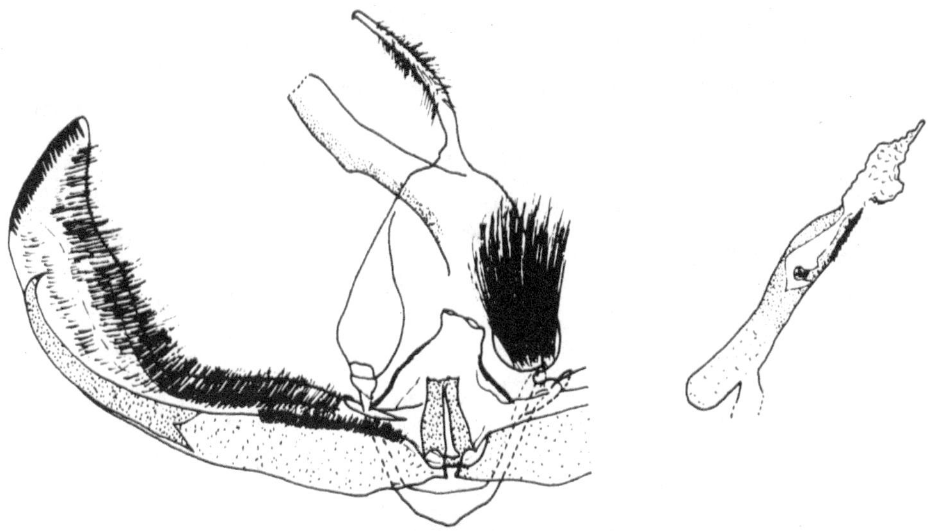

Fig. 22. *Argyrolepidia pamphilia* (STOLL). B. M. Noct. 8433

fold rather broad, occupying the proximal two-thirds of valva; harpe placed slightly before middle of valva, rather slender, reaching to the last fifth of valva. Aedeagus shorter than half of valva, rather slender, straight, bearing a terminal field of minute cornuti; fultura inferior shaped as a parallelogram. Saccus short, slightly angulate.

Distribution: Indonesia to New Guinea, Bismarck Archipelago, Solomon Islands, and Australia.

Key to the species

1 Head and thorax not yellow above . . . . . . . . . . . . 2
1′ Head and thorax, or part of thorax alone, yellow above . . . . . . . . . . . . 13
2 Head and thorax chocolate or deep red brown . . . . . . . . . . . . 4
2′ Head and thorax not chocolate or red brown . . . . . . . . . . . . 3
3 Head and thorax greyish brown . . . . . . . . . . . . *A. mutans*, p. 109
3′ Head and thorax blackish brown or black, sometimes glossed with blue . . . . . . . . . . 7
4 Hind wings with a large white spot . . . . . . . . . . . . 5
4′ Hind wings with large silvery or metallic blue spots . . . . . . . . . . . . 6
5 Hind wings with a large, irregularly elliptical median spot; cilia spotted with white . . . *A. unimacula*, p. 115
5′ Hind wings with some blue at the base, and a large elliptical median spot, edged with blue scales proximally; cilia tipped with white in the fore protion . . . . . *A. concisa*, p. 113
6 Hind wings black with blue reflections, without markings beyond the metallic blue median patch . . . . . . . . . . . . *A. palaea*, p. 109
6′ Hind wings black, with a lunulate white patch beyond the silvery blue spot, and with two small spots below veins 3 and 4 . . . . . . . . . . . . *A. lunaris*, p. 109
7 Fore wings without white spots below and beyond cell . . . . . . . . . . . . 8
7′ Fore wings with white spots below and beyond cell . . . . . . . . . . . . 12
8 Hind wings black with blue or green basal reflections, otherwise unspotted . . . . . . . . 9
8′ Hind wings black with pale markings . . . . . . . . . . . . 10
9 Abdomen black, ringed with whitish . . . . . . . . . . . . *A. novaehiberniae*, p. 107
9′ Abdomen bluish grey with the tip orange . . . . . . . . . . . . *A. aethrias*, p. 108

10 Abdomen black, ringed with pale . . . . . . . . . . . . . . . . . . . . . . . . . . . . . . . . . 11
10′ Abdomen orange, with the basal urite black, and the next one banded with black . . . . *A. aurea*, p. 116
11 Hind wings black with blue reflections, and a large rounded white patch . . . . . . . . . *A. resplendens*, p. 110
11′ Hind wings black, with silvery blue markings, sometimes obsolescent . *A. pamphilia*, p. 106
12 Abdomen with a black patch on the orange anal tuft. . . . . . . . . . . . . *A. fracta*, p. 110
12′ Abdomen with the orange anal tuft unspotted . . . . . . . . . . . . . . . *A. aequalis*, p. 111
13 Head and thorax orange, spotted with black . . . . . . . . . . . . . . . . . . . . . . . . 14
13′ Head black; meso- and metathorax pale orange . . . . . . . . . . . . . . . *A. stevensi*, p. 119
14 Fore wings with pale streaks on veins in terminal area . . . . . . . . . . . . . . . . . . . 15
14′ Fore wings without pale streaks on veins in terminal area . . . . . . . . . . . . . . . . 17
15 Fore wings with an oblique white postmedian band . . . . . . . . . . . . *A. megisto*, p. 117
15′ Fore wings with a white spot at lower cell-angle . . . . . . . . . . . . . . . . . . . . . . 16
16 Hind wings unspotted bluish black . . . . . . . . . . . . . . . . . . . . . . *A. restricta*, p. 118
16′ Hind wings with the basal area white . . . . . . . . . . . . . . . . . . . . . *A. comma*, p. 118
17 Collar edged with orange; fore wings with a white spot in cell . . . . . . . *A. goldiei*, p. 114
17′ Collar edged with white; fore wings without white spot in cell . . . . *A. inconspicua*, p. 115

## Argyrolepidia pamphilia (STOLL)

1782 *Geometra pamphilia* STOLL, Pap. Exot., **4**:153; pl. 368G. – Terra typica: Not designated (Type not designated).
1874 *Agarista pamphilia* – BOISDUVAL, Rev. Mag. Zool., (3) **2**:77.
1892 *Phalaenoides pamphilia* – KIRBY, Syn. Cat. Lepidopt. Het.:24.
1901 *Argyrolepidia pamphilia* – HAMPSON, Cat. Lepidopt. Phal. Brit. Mus., **3**:550 (♂).
1901 *Argyrolepidia pamphila* HAMPSON (ex errore pro *pamphilia*), Cat. Lepidopt. Phal. Brit. Mus. **3**:551; fig. 243 (♂).
1920 *Odice pamphila* – HAMPSON, Cat. Lepidopt. Phal. Brit. Mus., Suppl. **2**:544.

Diagnosis: Body and wings black with blue reflections; palpi fringed with orange hair, and with some white in front; first antennal joint, sides of frons, spots on occiput, throat, edges of collar and spots on tegulae white; fore coxae and femora with orange pilosity; markings on tibiae and tarsi white; abdomen with bluish white segmental bands, confluent below, and the anal tuft orange. Fore wings with a square white spot in middle of cell; other markings silvery or metallic blue, viz., a slight streak below base of vein 1; four subbasal spots below costa and cell; antemedial dots below costa, in cell and above vein 1; oblique medial spots between veins 2 and 7, with a spot above them below costa; a curved postmedial maculate line; and a subterminal series of spots; cilia white apically and tornally. Hind wings with silvery blue markings, sometimes obsolescent, viz., a spot at the lower cell-angle, and postmedial spots between veins 2 and 4 conjoined to a diffuse terminal row; cilia white. Length of fore wing 19 mm.

Distribution: Moluccas.

### Argyrolepidia pamphilia pamphilia (STOLL)

1782 *Geometra eusebia* STOLL, Papill. Exot., **4**:155–156; pl. 369G, H. – Terra typica: Amboina (Type not designated).
1864 *Agarista micacea* WALKER, List. Lepidopt. Het. Brit. Mus., **31**:48 (♂). – Terra typica: Ceram (Type, ♂, in Museum Oxford University).

1875 *Phalaenoides ceramensis* KIRBY, Cist. Entomol., 1:346. – Terra typica: Ceram (Type, ♂, in British Museum).
1912 *Argyrolepidia pamphilia pamphilia* – JORDAN in SEITZ, Großschmetterl. Erde, 11:22; pl. 3c.

Diagnosis: As above.

Distribution: Ambon, Ceram.

### Argyrolepidia pamphilia ombirana JORDAN

1906 *Argyrolepidia pamphilia ombiranus* JORDAN, Novit. Zool., 13:161 (♀). – Terra typica: Great Obi Island (Type, ♀, in British Museum).
1920 *Odice ombirana* – HAMPSON, Cat. Lepidopt. Phal. Brit. Mus., Suppl. 2:544.

Diagnosis: Blue markings very much reduced; discal white band on fore wings 3 – 4 mm broad.

Distribution: Obi Major Island.

### Argyrolepidia pamphilia madina JORDAN

1912 *Argyrolepidia pamphilia madina* JORDAN in SEITZ, Großschmetterl. Erde, 11:22 (♂). – Terra typica: Buru Island (Type, ♂, in British Museum).

Diagnosis: Blue markings well developed; white band on fore wings narrow; blue terminal spots on hind wings form a band, slightly broadening towards the tornus, running from the apex or from the subcostal vein to the lower media.

Distribution: Buru Island.

### Argyrolepidia novaehiberniae (BOISDUVAL)

1832 *Agarista novae-hiberniae* BOISDUVAL, Voyage Astrolabe, Lepidopt.:178. – Terra typica: New Ireland (Type, ♂, in British Museum).
1892 *Phalaenoides novae-hiberniae* – KIRBY, Syn. Cat. Lepidopt. Het.:23.
1901 *Argyrolepidia novae-hiberniae* – HAMPSON, Cat. Lepidopt. Phal. Brit. Mus., 3:550; pl. 52, fig. 9 (♂).

Diagnosis: Body and wings black; first palpal joint in front, first and second behind orange, second and third with a few white scales; frons and vertex white laterally; collar with a white dot at middle; tegulae and thorax with some white hair; pectus and legs black and white; fore coxae and fringes on femora orange; abdomen ringed with white except at the extremity; anal urite orange. Fore wings with a subbasal and an antemedial series of small metallic blue dots and a short streak on base of vein 1; a small white spot in middle of cell, with a blue spot above it in cell; spots below base of vein 2 and on middle of dorsum; a series of six white dots beyond cell between veins 2 and 7, with a small blue spot above them on costa, and a series distally from them, incurved towards the dorsum;

a subterminal series of dots; cilia white apically. Hind wings metallic blue on basal half except on costal area; outer edge of the blue area sinuous; a few subterminal blue scales on veins 2 and 3; cilia white, spotted with black. Length of fore wing 17 mm.

Distribution: Bismarck Archipelago.

### Argyrolepidia novaehiberniae novaehiberniae (BOISDUVAL)

Diagnosis: Mid coxae with some orange; hind wings glossy metallic blue from the base to beyond the cell-apex.

Distribution: New Hanover, New Ireland.

### Argyrolepidia novaehiberniae mesotis JORDAN

1926 *Argyrolepidia novaehiberniae mesotis* JORDAN, Novit. Zool., **33**:362 (♂, ♀). – Terra typica: New Britain (Types, ♂, ♀, in British Museum).

Diagnosis: Mid coxae without orange at the extremity; fore wings with the white discal band narrower, and the white apical spots smaller, especially in female; the blue discal spots and those between cell and dorsum also smaller; hind wings with a large, transparent white spot outside the cell-end, shaded with blue.

Distribution: New Britain.

### **Argyrolepidia aethrias** TURNER

1877 *Agarista novae-hiberniae* BUTLER, Illustr. Het. Brit. Mus., **1**:8; pl. 4, fig. 5 (♂).
1908 *Argyrolepidia aethrias* TURNER, Trans. R. Soc. South Australia, Adelaide, **1908**:55. – Terra typica: Northwest Australia (Type, ♂, in Australian Museum, Sydney).
1910 *Idalima cyanobasis* HAMPSON, Cat. Lepidopt. Phal. Brit. Mus., **9**:460 (nomen novum pro *Agarista novae-hiberniae* BUTLER, nec BOISDUVAL).
1912 *Idalima aethrias* – STRAND, Lepidopt. Cat., **5**:57.

Diagnosis: Head and thorax black; 1st and 2nd palpal joints yellow with a few black scales; sides of frons, vertex, collar, tegulae and thorax above striped with pale yellow; pectus with pale yellow pilosity; legs black striped with white; mid tibiae above orange at middle; abdomen grey above; anal tuft orange with a black spot; underside white at base, orange towards the extremity. Fore wings black with greenish yellow basal spots; two short subbasal subcostal lines, a wedge-shaped patch below cell, its tip connected by a bar with a streak on base of dorsum; a medial band from cell to dorsum with an oblique mark beyond it from the submedian fold to dorsum; purplish blue spots in middle and end of cell; a yellow band beyond cell from costa to vein 2, oblique and sinuous to vein 5, then bent outwards and with three strong dentations on distal side, and with some purplish blue scales running obliquely from its dorsal extremity to the oblique dorsal mark; a subterminal row of purplish blue dots on veins, that on vein 5 more proximal; cilia white apically and above the tornus. Hind wings black; basal half suffused with greenish yellow except on costal area; cilia chequered with white and black. Length of fore wing 19 mm.

D i s t r i b u t i o n: North West Australia; Northern Queensland.

### **Argyrolepidia palaea** ROTHSCHILD & JORDAN

1905 *Argyrolepidia palaea* ROTHSCHILD & JORDAN, Novit. Zool., **12**:473 (♀). – T e r r a t y p i c a: Angabunga River, Papua (Type, ♀, in British Museum).
1920 *Odice palaea* – HAMPSON, Cat. Lepidopt. Phal. Brit. Mus., Suppl. **2**:544.

D i a g n o s i s: Head and thorax deep red brown; basal joint of the antennae white in front; palpi orange at base; second joint with some white scales in front and at the extremity; a white postocular line; pectus and femora orange red; tibiae and tarsi dark brown ringed with white; abdomen dark brown with bluish white segmental lines. Fore wings deep red brown; an indistinct pale oblique line in cell-middle and a pale discocellular bar with a very indistinct oblique band from it to vein 2. Hind wings black with blue reflections, and a large oblique metallic blue patch; cilia with a white spot at the submedian fold. Length of fore wing 22 mm.

D i s t r i b u t i o n: Papua (Angabunga River = terra typica).

### **Argyrolepidia mutans** (ROTHSCHILD)

1901 *Phalaenoides mutans* ROTHSCHILD, Novit. Zool., **8**:220; pl. 9, fig. 4 (♂). – T e r r a t y p i c a: Aroa River, Papua (Type, ♂, in British Museum).
1912 *Argyrolepidia mutans* – JORDAN in SEITZ, Großschmetterl. Erde, **11**:23.
1920 *Odice mutans* – HAMPSON, Cat. Lepidopt. Phal. Brit. Mus., Suppl. **2**:547.

D i a g n o s i s: Head and thorax greyish brown; palpi in front and pilosity of pectus and legs orange yellow; abdomen fuscous black; underside yellowish except towards the extremity. Fore wings uniformly dark brown with a cupreous tinge; a few blue basal points; a narrow, indistinct discocellular band. Hind wings black; a large white patch in and below cell-extremity, edged with metallic blue, diffused towards the base of the wing. Length of fore wing 20 mm.

D i s t r i b u t i o n: Papua (Aroa River = terra typica).

### **Argyrolepidia lunaris** ROTHSCHILD & JORDAN

1905 *Argyrolepidia lunaris* ROTHSCHILD & JORDAN, Novit. Zool., **12**:473 (♂). – T e r r a t y p i c a: Angabunga River, New Guinea (Type, ♂, in British Museum).
1920 *Odice lunaris* – HAMPSON, Cat. Lepidopt. Phal. Brit. Mus. Suppl. **2**:543.

D i a g n o s i s: Head and thorax chocolate brown, tinged with grey; palpi black with the base orange; 2nd joint with some white hair mixed; frons marked with white; genae white; pectus and pilosity of femora orange; tibiae marked with white; tarsi ringed with white; abdomen chocolate brown, banded with bluish;

underside banded with white; anal tuft orange. Fore wings chocolate brown; a white bar across middle of cell with a few brown scales; an oblique postmedial series of white spots in interspaces, sprinkled with brown, running from costa to vein 2, incurved at vein 5, then excurved; metallic blue markings, viz., a subbasal subcostal striga; a subcellular dot; an antemedial series of points from costa to vein 1, incurved in cell, then excurved; distally, a point in interval I; an oblique line of scales below the submedian fold; a curved striga above dorsum beyond the middle; a point beyond the postmedial band in interval VII; subterminal points on veins 1, 2 and 3; cilia with a white subapical patch. Hind wings black, with a large silvery blue patch in and beyond cell; a streak on vein 1 to beyond middle; a white lunulate patch beyond cell, near the distal edge of the blue patch; small spots below veins 3 and 4; cilia white subapically and at vein 1. Length of fore wing 23 mm.

Distribution: Papua (Angabunga River = terra typica).

### **Argyrolepidia resplendens** (ROTHSCHILD & JORDAN)

1903 *Phalaenoides resplendens* ROTHSCHILD & JORDAN, Novit. Zool., **10**:487; pl. 12, fig. 37 (♂). – Terra typica: Upper Aroa River, Papua (Type, ♂, in British Museum).
1912 *Argyrolepidia resplendens* JORDAN in SEITZ, Großschmetterl. Erde, **11**:23; pl. 3b.

Diagnosis: Head black; 1st palpal joint and sides of 2nd fringed with grey; frons, vertex and genae grey laterally; thorax blue black; shoulders and collar in front with some orange; pectus orange; abdomen blue black with bands and underside greenish grey; anal tuft orange at the extremity. Fore wings black with blue reflections; veins of the distal half with fine blue grey streaks; an oblique white postmedial band from below costa to vein 2 near the termen, with some blue grey above it below costa, and two slight blue grey lines beyond it before termen. Hind wings with the termen slightly excurved at middle, black with brilliant blue reflections, and a large rounded white patch. Length of fore wing 26 mm.

Distribution: Papua (Aroa River).

### **Argyrolepidia fracta** (ROTHSCHILD)

1899 *Phalaenoides fracta* ROTHSCHILD, Novit. Zool., **6**:432 (♂, ♀). – Terra typica: Kapaur, New Guinea (Types, ♂, ♀, in British Museum).
1901 *Argyrolepidia fracta* HAMPSON, Cat. Lepidopt. Phal. Brit. Mus., **3**:552; pl. 52, fig. 11.

Diagnosis: Head, body and fore wings black brown; head and thorax mixed with white hair; second palpal joint mostly yellow; sides of frons white; neck yellowish; coxae and fore femora fringed with yellow hair; abdomen banded dorsally with bluish white; underside white; anal tuft orange with a black patch. Fore wings with a square white spot in middle of cell; a sinuous series of five spots between veins 2 and 7 beyond cell, with a series of blue spots beyond them; other markings blue, viz., two dots below base of costa; two on the media; larger spots at base of dorsum and in interspace I before middle; a postmedial spot on vein 1; a subterminal series of dots. Hind wings black, with a blue streak below base of cell; a medial white spot below cell; a spot beyond cell traversed by a blue streak on vein 4; cilia chequered black and white. Length of fore wing 18 mm.

Distribution: New Guinea; Northern Queensland.

### Argyrolepidia aequalis (WALKER)

1864 *Agarista aequalis* WALKER, List Lepidopt. Het. Brit. Mus., **31**:47 (♀). – Terra typica: Aru Island (Type, ♀, in Museum Oxford University).
1892 *Phalaenoides aequalis* – KIRBY, Syn. Cat. Lepidopt. Het.:23.
1892 *Rothia aequalis* – SWINHOE, Cat. Het. Mus. Oxon:159.
1901 *Argyrolepidia aequalis* – HAMPSON, Cat. Lepidopt. Phal. Brit. Mus., **3**:552.

Diagnosis: Head, body and wings blackish brown tinged with blue; palpi with orange and white pilosity; first joint of antennae, sides of frons, spots on vertex, edge of collar and spots on tegulae white; pectus in front and pilosity of fore coxae and femora orange; remainder of pectus and stripes on legs white; abdomen banded with white, the bands confluent on the underside; extremity of abdomen orange. Fore wings with a white spot in middle of cell; an oblique maculate band beyond cell from interspace II to vein 7; remainder of markings blue, viz., a short streak below base of vein 1; a basal spot; subbasal spots below costa and cell, followed by spots below costa and in cell, and by antemedial spots below costa and in interspace I; a spot below base of vein 2; a curved postmedial series of spots below costa and between dorsum and vein 4; a subterminal series of spots, obsolescent towards costa; cilia white apically and near tornus. Hind wings with brilliant blue sheen; large white spots below end of cell and beyond the discocellulars; cilia white towards apex and with series of white spots. Length of forewing 21 – 24 mm.

Distribution: Aroe Islands to Solomon Islands; Queensland.

### Argyrolepidia aequalis aequalis (WALKER)

1894 *Phalaenoides dohertyi* OBERTHÜR, Études Entomol., **19**:24; pl. 5, fig. 28 (♂). – Terra typica: New Guinea (Type, ♂, in British Museum).
1912 *Argyrolepidia aequalis aequalis* – JORDAN in SEITZ, Großschmetterl. Erde, **11**:23; pl. 3c.

Diagnosis: Hind wings with the white spots conjoined in female, and sometimes in male.

Distribution: New Guinea, Aroe Islands, Fergusson Island.

### Argyrolepidia aequalis caeruleotincta (LUCAS)

1891 *Agarista caeruleotincta* LUCAS, Proc. Linn. Soc. New South Wales, Sydney, (2) **6**:303 (♂). – Terra typica: Queensland (Type, ♂, in South Australian Museum, Adelaide).

1901 *Argyrolepidia caeruleotincta* – HAMPSON, Cat. Lepidopt. Phal. Brit. Mus., 3:553; pl. 52, fig. 12.
1912 *Argyrolepidia aequalis caeruleotincta* – JORDAN in SEITZ, Großschmetterl. Erde, **11**: 23; pl. 3b.

Diagnosis: General colouration paler brown; markings on thorax and fore wings much reduced; oblique maculate band on fore wings reaching the costa; white spots on hind wings always separated.

Distribution: Queensland.

## Argyrolepidia aequalis capiens JORDAN

1912 *Argyrolepidia aequalis capiens* JORDAN, Novit. Zool., **11**:446 (♂, ♀). – Terra typica: Treasury Island (Types, ♂, ♀, in British Museum).
1920 *Odice capiens* – HAMPSON, Cat. Lep. Phal., Suppl. **2**:546; pl. 70, fig. 14.

Diagnosis: Palpi, legs and underside of abdomen less mixed with white scales; hind wings with the lower white spot sometimes much reduced in male, small in female.

Distribution: Treasure Island.

## Argyrolepidia aequalis integra JORDAN

1904 *Argyrolepidia aequalis integra* JORDAN, Novit. Zool., **11**:446 (♂, ♀). – Terra typica: Choiseul Island, Solomon Islands (Types, ♂, ♀, in British Museum).
1920 *Odice integra* – HAMPSON, Cat. Lepidopt. Phal. Brit. Mus., Suppl. **2**:545; pl. 70, fig. 6.

Diagnosis: Fore wings with the oblique maculate band 2 – 3 mm wide in male, 4 – 5 mm in female, followed by a mostly continous blue band; hind wings with the white spot large, as broad as the terminal black area in male, and nearly twice as broad in female.

Distribution: Solomon Islands (Bougainville, Choiseul, Santa Isabel, Rendova, Guizo, Vella Lavella, Kolombangara).

## Argyrolepidia aequalis salomonis (ROTHSCHILD)

1899 *Phalaenoides aequalis salomonis* ROTHSCHILD, Novit. Zool., **6**:432 (♀). – Terra typica: Tulagi, Guadalcanal (Type, ♀, in British Museum).
1901 *Argyrolepidia salomonis* – HAMPSON, Cat. Lepidopt. Phal. Brit. Mus., 3:553; pl. 52, fig. 19.
1912 *Argyrolepidia aequalis salomonis* – JORDAN in SEITZ, Großschmetterl. Erde, **11**:23; pl. 2g.

Diagnosis: Nearest to *A. a. capiens*, but the white spots on hind wings somewhat larger, and with broader connection with the media, and the basal area more strongly sprinkled with blue.

Distribution: Guadalcanal Island (Tulagi = terra typica).

### Argyrolepidia aequalis repens JORDAN

1939 *Argyrolepidia aequalis repens* JORDAN, Novit. Zool., **41**:249–250 (♀). – Terra typica: Tuelo, Halmahera Island (Type, ♀, in British Museum).

Diagnosis: Fore wings with the whitish blue spots reduced; a minute white dot in cell; hind wings with a small white spot outside the discocellulars. Length of fore wing 24 mm.

Remark: Probably not conspecific with *A. aequalis* because of the disjunct distribution; no forms of *A. aequalis* are known to occur in southern and eastern Moluccas.

Distribution: Halmaheira (Toeelo = terra typica).

## Argyrolepidia concisa JORDAN

1912 *Argyrolepidia concisa* JORDAN in SEITZ, Großschmetterl. Erde, **11**:23 (♂). – Terra typica: Oetakwa, Snow Mountains, New Guinea (Type, ♂, in British Museum).
1920 *Odice concisa* – HAMPSON, Cat. Lepidopt. Phal. Brit. Mus., Suppl. **2**:545; pl. 70, fig. 7.

Diagnosis: Head, thorax and fore wings deep chocolate brown; palpi with orange pilosity in front and behind towards the base, and with some white hair in front of the second joint; sides of head white; vertex, collar and tegulae mixed with bluish white hairs; antennae black; pectus orange; legs blackish brown; tibiae and tarsi ringed with white; abdomen blackish brown with bluish white segmental lines; underside white except urites VII and VIII; anal tuft orange. Fore wings with the markings silvery blue, viz., some basal dots; a few scales above base of dorsum; minute acute dots below costa, in cell and in interval I; a short postmedial streak below costa; a narrow oblique maculate band between veins 2 and 7; cilia tipped with white apically. Hind wings slightly darker; base with some silvery blue scales; a large elliptical white, slightly transparent spot in middle of wing, slightly edged with silvery blue scales proximally; cilia tipped with white from near apex to discal fold. Length of fore wing 21 mm.

Distribution: Irian.

### Argyrolepidia concisa concisa (JORDAN)

Diagnosis: As above.

Distribution: Irian (Snow Mountains, Oetakwa River).

### Argyrolepidia concisa stilbalis JORDAN

1930 *Argyrolepidia concisa stilbalis* JORDAN, Novit. Zool., **35**:287. – Terra typica: Papua. Hydrographer Mts. (Type, ♂, in British Museum).

Diagnosis: Forewings with the discal band broader; hind wings with the white patch larger and bordered with blue all round.

Distribution: Papua.

## Argyrolepidia goldiei (DRUCE)

1894 *Agarista goldiei* DRUCE, Ann. Mag. Nat. Hist., London, (6) **14**:21 (♂). – Terra typica: Port Moresby, New Guinea (Type, ♂, in British Museum).
1896 *Phalaenoides goldiei* – ROTHSCHILD, Novit. Zool., **3**:49.
1912 *Argyrolepidia goldiei* – JORDAN in SEITZ, Großschmetterl. Erde, **11**:23.

Diagnosis: Head, body and wings black; palpi orange below at base, white towards the tip of the second joint; sides of frons and dots on occiput white; collar, and fore femora and tibiae orange; pectus mixed with white hair; abdomen whitish at the base; anal tuft orange. Fore wings with a white dot in middle of cell; an oblique white postmedial band between veins 2 and 7. Hind wings with a broad white medial band from the anal margin to the subcostal vein, it distal edge minutely dentate; cilia white apically. Length of fore wing 21 mm.

Distribution: New Guinea.

### Argyrolepidia goldiei goldiei (DRUCE)

1912 *Argyrolepidia goldiei goldiei* – JORDAN in SEITZ, Großschmetterl. Erde, **11**.

Diagnosis: As above.

Distribution: Papua (Port Moresby [terra typica], Aroa River region).

### Argyrolepidia goldiei figurata JORDAN

1912 *Argyrolepidia goldiei figurata* JORDAN in SEITZ, Großschmetterl. Erde, **11**:23–24; pl. 3a (♂). – Terra typica: Humboldt Bay, New Guinea (Type, ♂, in British Museum).
1920 *Phalaenoides goldiei* subsp. *figurata* – HAMPSON, Cat. Lepidopt. Phal. Brit. Mus., Suppl. **2**:548.

Diagnosis: Abdomen not whitish at the base above; fore wings with some blue dots at base and distally of the white postmedial band; a white dot in cell in male; hind wings with the white band not reaching the anal margin.

Distribution: Papua (Humboldt Bay [terra typica], Northeast Papua).

### Argyrolepidia goldiei perisa JORDAN

1912 *Argyrolepidia goldiei perisa* – JORDAN in SEITZ, Großschmetterl. Erde, **11**:24 (♂). – Terra typica: Milne Bay, New Guinea (Type, ♂, in British Museum).
1920 *Phalaenoides perisa* – HAMPSON, Cat. Lepidopt. Phal. Brit. Mus., Suppl. **2**:548; pl. 70, fig. 16.

Diagnosis: Abdomen mixed with white hair above on urites II and III in male; anal tuft with a black median spot; fore wings with blue dots at the base only.

Distribution: Papua (Milne Bay = terra typica).

## Argyrolepidia goldiei cremor JORDAN

1912 *Argyrolepidia goldiei cremor* JORDAN in SEITZ, Großschmetterl. Erde, **11**:24 (♂). – Terra typica: Mount Goliath, New Guinea (Type, ♂, in British Museum).
1920 *Phalaenoides cremor* – HAMPSON, Cat. Lepidopt. Phal. Brit. Mus., Suppl. **2**:548; pl. 70, fig. 17.

Diagnosis: Fore and hind coxae with pale yellow pilosity; tegulae without white spot; urites II–IV yellowish white, urites V–VII without white segmental lines; fore wings with a bluish white costal dot at base, but without blue dots; hind wings with the discal band cream coloured, reaching the anal margin where it is tinged with blue; cilia tipped with white towards the apex.

Distribution: Irian (Mount Goliath).

## Argyrolepidia goldiei similis HULSTAERT

1923 *Argyrolepidia goldiei similis* HULSTAERT, Ann. Mag. Nat. Hist., London, (9) **11**:188 (♂). – Terra typica: Merauke, New Guinea (Type, ♂, in British Museum).

Diagnosis: Collar edged with bluish white; abdomen black with white segmental bands; fore wings with blue basal and postdiscal spots; a white spot in cell; the white postmedial band widest between veins 3 and 4; hind wings with a white patch glossed with blue, not reaching the anal margin and incised on vein 2.

Distribution: Irian (Merauke, about 70 km from the Papuan boundary, on the south coast).

## **Argyrolepidia inconspicua** (ROTHSCHILD)

1896 *Phalaenoides inconspicua* ROTHSCHILD, Novit. Zool., **3**:49; pl. 15, fig. 30 (♂). – Terra typica: Humboldt Bay, New Guinea (Type, ♂, in British Museum).
1912 *Argyrolepidia inconspicua* – JORDAN in SEITZ, Großschmetterl. Erde. **11**:24.

Diagnosis: Head, body and wings black; palpi with the second joint with orange hair at base, and white in front; head mixed with white; collar slightly edged with white; fore coxae and fringes of femora orange; mid and hind coxae with white pilosity; abdomen with white segmental lines and underside; anal tuft greyish yellow; last sternite blackish in female, tipped with orange. Fore wings with faint bluish reflections; a white band beyond the cell, from interval I to vein 7, angled at vein 5. Hind wings with a very broad white band in middle, from the anal margin to the subcostal vein, twice as broad at middle as the black marginal band; its distal edge slightly dentate. Length of fore wing 22 – 27 mm.

Distribution: New Guinea (Humbold Bay, Kapaur).

## **Argyrolepidia unimacula** LOWER

1903 *Argyrolepidia unimacula* LOWER, Trans. R. Soc. South Australia, Adelaide, **27**:29 (♀). – Terra typica: Mackay, Queensland (Type, ♀, in South Australian Museum).
1920 *Odice unimacula* – HAMPSON, Cat. Lepidopt. Phal. Brit. Mus., Suppl. **2**:546; pl. 70, fig. 15.

D i a g n o s i s: Head, thorax and fore wings chocolate brown; 2nd palpal joint white in front; spot behind the base of antenna and a postocular line white; pectus yellowish white; legs banded with dark brown and white; abdomen (wanting in the single specimen known). Fore wings with a slight whitish discocellular lunule; an oblique bluish white band from costa at two-thirds, narrowing to a point at tornus, and with the edges somewhat sinuous. Hind wings blackish brown; a large, irregularly elliptical white spot in middle; cilia dotted with white. Length of fore wing 18 mm.

D i s t r i b u t i o n: Queensland (Mackay = terra typica).

## **Argyrolepidia aurea** JORDAN

1903 *Argyrolepidia aurea* JORDAN, Novit. Zool. **10**:315; pl. 12, fig. 21 (♂, ♀). – T e r r a t y p i c a: Aroa River, New Guinea (Type, ♂, in British Museum).
1920 *Odice aurea* – HAMPSON, Cat. Lepidopt. Phal. Brit. Mus., Suppl. **2**:547.

D i a g n o s i s: Head and thorax bluish black; base and sides of the second palpal joint white; sides of frons, vertex and genae white; collar slightly edged with white; pectus and pilosity of fore femora yellowish white; abdomen orange; first urite black, second with a black band; underside black, except the last two urites, banded with white and with a white ventral stripe except on the last three urites. Fore wings with bluish reflections; a broad deep orange band from costa beyond middle to tornus where it is dilated above dorsum. Hind wings with blue reflections, and with an orange band from below costa beyond middle, broadening towards anal margin where it extends from near base to tornus, slightly narrower in female. Length of forewing 22 mm.

D i s t r i b u t i o n: New Guinea.

### A r g y r o l e p i d i a a u r e a a u r e a JORDAN

1912 *Argyrolepidia aurea aurea* – JORDAN in SEITZ, Großschmetterl. Erde, **11**:24; pl. 3a.

D i a g n o s i s: As above, with the orange band on fore wings but slightly extended along tornus.

D i s t r i b u t i o n: Mount Owen Stanley (terra typica).

### A r g y r o l e p i d i a a u r e a f e r v i d a JORDAN

1912 *Argyrolepidia aurea fervida* JORDAN in SEITZ, Großschmetterl. Erde, **11**:24 (♂, ♀). – T e r r a t y p i c a: Lower Oetakwa River, New Guinea (Type, ♂, in British Museum).

D i a g n o s i s: The three basal urites nearly black above; base of urite IV black; urites V and VI blackish subbasally; urite VII blackish at middle.

D i s t r i b u t i o n: Lower Utakwa River (terra typica).

### Argyrolepidia aurea angustifasciata JOICEY & TALBOT

1922 *Argyrolepidia aurea angustifasciata* JOICEY & TALBOT, Bull. Hill Mus., Witley, London, 1:301 (♂). – Terra typica: Wangaar, New Guinea (Type, ♂, in British Museum).

Diagnosis: The four last urites scaled with orange on the underside; fore wings with the orange band much narrower, reduced proximally; hind wings straight distally, and curved posteriorly.

Distribution: Irian (Wangaar = terra typica).

## Argyrolepidia megisto (BOISDUVAL)

1832 *Agarista megisto* BOISDUVAL, Voyage Astrolabe, Lepidopt.: 179; pl. 5, fig. 5. – Terra typica: Dorey, New Guinea (Type, ♂, in British Museum).
1864 *Vitessa pyraliformis* WALKER, List. Lepidopt. Het. Brit. Mus., 31:220 (♀). – Terra typica: Dorey, New Guinea (Type, ♀, in Museum Oxford University).
1889 *Agarista neurogramma* MEYRICK, Trans. Entomol., Soc. London, **1889**:468 (♂). – Terra typica: Mount Obree, New Guinea (Type, ♂, in British Museum).
1901 *Phalaenoides megisto* – HAMPSON, Cat. Lepidopt. Brit. Mus., 3:557; pl. 52, fig. 6 (♀).
1900 *Phalaenoides basiplaga* ROTHSCHILD, Novit. Zool., 7:275; pl. 5, fig. 5 (♀). – Terra typica: Papua, Milne Bay (Type, ♀, in British Museum).
1901 *Phalaenoides neurogramma* – HAMPSON, Cat. Lepidopt. Phal. Brit. Mus., 3:556; pl. 52, fig. 8.
1912 *Argyrolepidia megisto* – JORDAN in SEITZ, Großschmetterl. Erde, 11:24.
1912 *Argyrolepidia megisto* forma *neurogramma* – JORDAN in SEITZ, Großschmetterl. Erde, 11:24; pl. 3b.
1912 *Argyrolepidia megisto* forma *basiplaga* – JORDAN in SEITZ, Großschmetterl. Erde, 11:24; pl. 3b.
1912 *Argyrolepidia megisto* forma *intermedia* – JORDAN in SEITZ, Großschmetterl. Erde, 11:24.

Diagnosis: Head and thorax orange; palpi with some black hair; 2nd joint behind and 3rd black; antennae black; collar and tegulae spotted with black; legs black with whitish streaks; abdomen bluish black; last urite orange. Fore wings steely blue black, browner in dorsal area, sometimes more or less broadly bluish grey in medial area (forma "*neurogramma*"); a subbasal pale (greyish or yellowish white) line forming a patch on dorsum; a postmedial yellowish white band from vein 7 just beyond cell to vein 1 near tornus. Hind wings blackish brown with blue reflections, without (forma "*megisto*", forma "*neurogramma*") or with (forma "*basiplaga*") white spot, sometimes irrorated with black (forma "*intermedia*"); cilia grey except towards apex. Length of fore wing 20 – 26 mm.

Distribution: Irian; Aroe Islands.

### Argyrolepidia megisto megisto (BOISDUVAL)

1912 *Argyrolepidia megisto* forma *megisto* – JORDAN in SEITZ, Großschmetterl. Erde, 11:24; pl. 3b.

Diagnosis: As above.

Distribution: Irian.

### Argyrolepidia megisto cissia CLENCH

1953 *Argyrolepidia megisto cissia* CLENCH, Ann. Carnegie Mus., Pittsburgh, **33** (6):142. – Terra typica: Aru Island (Type, ♂, in Carnegie Museum, Pittsburgh).

Diagnosis: Differs from the nymotypical race in a narrow white cell-streak; white bases of the radius and the media (or the cubitalis) fusing with the postbasal transverse white band and forming a small square black cell-spot edged with white; postmedial white band on fore wings broader; hind wings uniform blackish brown; costa and cilia uniform white.

Distribution: Aroe Islands.

### Argyrolepidia comma TALBOT

1929 *Argyrolepidia comma* TALBOT, Bull. Hill Mus. Witley, London 3:91 (♀). – Terra typica: Nomnagihé, New Guinea (Type, ♀, in British Museum).

Diagnosis: Near *A. megisto*. Fore wings with heavily marked distal stripes of which there are eight, plus a thin streak on the submedian fold; postdiscal band as in *A. restricta*, but larger and broader, narrower costally, pointed just below costa, outcurved from vein 5 and reaching the submedia; cell-bar about comma-shaped, meeting a projection of postdiscal band at the origin of vein 2; discal band farther from basal band than in *A. megisto*, narrowing dorsally and not touching dorsum. Hind wings with a white proximal area from upper cell-margin to dorsum, and to well beyond the cell distally. Otherwise like *A. megisto*.

Remark: Talbot wondered whether *A. comma* be conspecific with *A. megisto* of which Nomnagihé is likewise the terra typica. The single female known might be an individual variant related to the form "*basiplaga*" of ROTHSCHILD (Milne Bay = terra typica).

Distribution: Irian (Nomnagihé = (terra typica).

### Argyrolepidia restricta (ROTHSCHILD)

1897 *Phalaenoides restrictus* (ROTHSCHILD), Novit. Zool., 4:311; pl. 7, fig. 8 (♀). – Terra typica: New Britain (Type, ♀, in British Museum).
1912 *Argyrolepidia restrictus* – JORDAN in SEITZ, Großschmetterl. Erde, **11**:24; pl. 3c.

Diagnosis: Head, collar and tegulae orange; antennae, 3rd palpal joint and patches on collar black; thorax black with an orange patch on metathorax; abdomen black with blue reflections; last urite orange. Fore wings black with brilliant blue reflections in basal two thirds; a whitish streak on base of dorsum; an oblique narrow yellowish white antemedial band not reaching the costa; a spot in cell-end; a patch beyond the lower cell-angle; veins of the terminal area streaked with whitish, except towards tornus. Hind wings bluish black, unspotted; cilia white except towards apex. Length of forewing 20 mm.

Distribution: Bismarck Archipelago (New Britain = terra typica, New Ireland, New Hanover).

## Argyrolepidia stevensi JORDAN

1938 *Argyrolepidia stevensi* JORDAN, Novit. Zool., **41**:133 (♀). – Terra typica: Morobe District, New Guinea (Type, ♀, in British Museum)

Diagnosis: Head black; antennae spotted with white above; 1st and 2nd palpal joints white beneath; 3rd as long as 2nd; sides of frons mixed with white; postocular region with white scales; pronotum edged with yellow in front, and with white laterally, posteriorly and medially; meso- and metathorax buffish, the latter with a pale yellow lateral tuft; legs black; femora clothed with yellow hair above; tibiae with some white scaling; tarsi with a few white scales; pectus and coxae orange yellow; abdomen black above with the urites edged with yellow; a large black basal tuft with some yellow at sides; anal tuft black with yellow hairs beneath; underside also banded with orange yellow. Fore wings black faintly shot with blue; a brownish grey spot beyond middle of cell; a brownish grey discal band about 2 mm wide, well beyond cell, ending at the lower media; metallic blue fine lines, one near the base, one halfway between the base and the lower media, one discal; cilia white apically. Hind wings deeper black; a pale orange spot below cell, close to the base of the lower media, with a larger, more or less rounded, somewhat oblique pale orange spot outside the discocellulars; cilia white apically. Length of fore wing 23 mm.

Distribution: Papua (Morobe Distict = terra typica; females only).

## Genus Phalaenoides LEWIN

(fig. 23)

1805 *Phalaenoides* LEWIN, Lepidopt. Insect. New South Wales:2. – Species typica: *Phalaenoides glycinae* LEWIN (by monotypy).

Diagnosis: Proboscis fully developed; antennae slightly depressed, without lateral compression, distinctly dilated towards the extremity which is blunt; palpi obliquely upturned; 1st and 2nd joints with long hair in front; 3rd porrect, naked, elongate, dilated at the extremity; frons with a truncate conical prominence, with raised rim at the extremity; head, thorax, femora fore and mid tibiae with long pilosity; hind tibiae with two pairs of spurs; base of abdomen with lateral tufts of long hair in male. Venation: in fore wings, vein 2 from nearly five sixths of cell; 3, 4, 5 well separated, equidistant; discocellulars faintly incurved; 6 from upper cell-angle; areole narrow; 7 and 8 + 9 at a point, from the extremity of areole; 10 from the upper margin of areole, near the extremity. Hind wings with vein 2 from about five sevenths of cell; 3, 4 and 6, 7 from angles of cell; discocellulars incurved; 5 obsolescent, from middle of discocellulars.

Male genitalia: Uncus very slender, arched, with a minute terminal hook. Tubus analis hardly sclerified. Fold of vinculum moderately broad, with a subbasal bunch of rather short stiff hair. Valva elongate, banana-shaped, distinctly broader in distal half; costa incurved, with a slight outcurving at middle; apex pointed; sacculus regularly outcurved; distal two-thirds of sacculus edged with stiff spicule-shaped hair; harpe as a process of the broad basal fold of sacculus, short, rather straight in distal two thirds. Aedeagus about half as long as the valva, slender, faintly S-shaped,

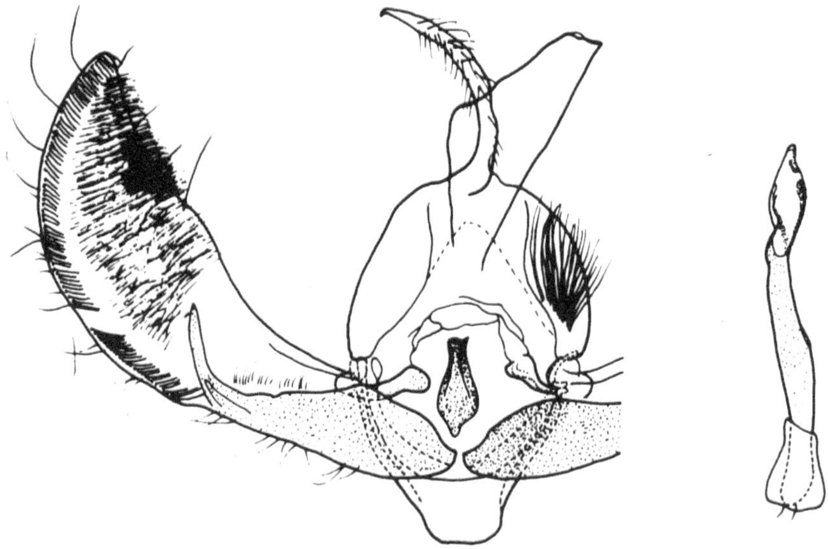

Fig. 23. *Phalaenoides glycinae* LEWIN. B. M. Noct. 8434

with a number of semi-microscopical terminal cornuti; fultura inferior narrowing distally. Saccus short, squarish.

Distribution: Australia.

## Key to the species

1 Fore wings with pale streaks on veins in terminal area . . . . . . . . . . . . . . . . . . . . 2
1' Fore wings without such streaks . . . . . . . . . . . . . . . . . . . . . . . . . *P. tristifica*, p. 121
2 Collar and tegulae yellow . . . . . . . . . . . . . . . . . . . . . . . . *P. thoracophora*, p. 122
2' Collar and tegulae black . . . . . . . . . . . . . . . . . . . . . . . . . . *P. glycinae*, p. 120

## Phalaenoides glycinae LEWIN

1805 *Phalaenoides glycinae* LEWIN, Lepidopt. Insect. New South Wales: 2; pl. 1 – Terra typica: Australia (Type not designated).
1836 *Agarista glycinae* BOISDUVAL, Spéc. Gén. Lépidopt., 1: pl. 14, fig. 3.

Diagnosis: Body and wings black; base of 1st and extremity of 2nd joint of palpi banded with yellow; sides of frons, vertex, collar, tegulae and middle of thorax with greenish-yellow stripes; throat yellow; pectus, femora and mid and hind tibiae with orange pilosity; fore tibiae fringed with yellow and black hair; tibiae and tarsi with white dots; terminal urite and a series of ventral bands yellow. Fore wings with two greenish-yellow basal dots; a greenish-yellow subbasal line from costa to submedian fold, from where there goes a streak nearly to the oblique greenish-yellow antemedial line, which is incurved to dorsum; a yellow bar across the cell; between it and the costa, a few greenish scales; an oblique yellow postdiscal band, rather sinuous, running from below costa just before two-thirds to near tornus, with some greenish scales beyond it between costa and vein 5, and below vein 1 in the distal half; a series of slight white subterminal nervural streaks, rather indistinct; cilia white apically. Hind wings with a yellow patch just beyond the cell, sometimes absent in male, absent in female; a series of pale yellow spots on termen; cilia yellow, forming together with the terminal spots a broad marginal line. Length of fore wing 23 – 26 mm.

Distribution: Australia (New South Wales, Victoria).

Early stages: **Larva** white to greenish yellow, especially on the underside; head marked with black; segments with numerous black lines, interrupted by dorsal and supraspiracular white lines; spiracles on red spots becoming large and square on the medial rings; eleventh ring with a broad dorso-lateral red band; hair sparse, black, not thickened terminally. On *Vitis*, sometimes becoming injurious, also on *Cissus*, *Gnaphalium luteoalbum*. Repulsive to birds; can emit through the mouth a greenish liquid. **Pupa** in a cocoon of agglutinate earth, in earth near the surface.

### Phalaenoides tristifica (HÜBNER)

1818 *Eutactis tristifica* HÜBNER, Zutr. Sammlg. exot. Schmetterl., **1**:28; figs. 165 and 166. – Terra typica: Australia (Type not designated).
1832 *Agarista lewini* BOISDUVAL, Voyage Astrolabe, **1** (Lépidopt):176. – Terra typica: Van Diemen's Land (Type, ♂, in British Museum).
1847 *Agarista ephyra* ANGAS, South Austr. Illustr.: pl. 37, fig. 20. – Terra typica: Van Diemen's Land (Type not designated).
1892 *Phalaenoides tristificus* – KIRBY, Syn. Cat. Lepidopt. Het.:22.
1901 *Phalaenoides tristifica* – HAMPSON, Cat. Lepidopt. Phal. Brit. Mus., **3**:559 (♂, ♀).

Diagnosis: Body and wings black; 2nd and 3rd joints of palpi and sides of frons with white dots; occiput, collar and tegulae spotted with greenish yellow; femora and mid and hind tibiae with orange pilosity; tibiae and tarsi with white points; abdomen with lateral series of whitish points; anal tuft and a ventral series of bands on the last four urites, orange. Fore wings with white points at base; a short subbasal ochreous band from costa; an antemedian ochreous band interrupted in the submedian fold; a square spot in middle of cell, with sometimes another below it; a postmedian band running from below costa to vein 2, excurved beyond the lower cell-angle; a spot nearer to the base of the dorsal area; two curved subterminal series of yellowish-white spots; cilia white apically and at tornus. Hind wings with or without a more or less distinct ochreous discal patch, with or without a postmedial series of white dots on veins; cilia spotted with white. Length of fore wing 19 – 23 mm.

Early stages: **Larva** pale yellow; head red marked with black; first and eleventh segments with red dorsal and lateral bands, with two lateral black spots; last pair of prolegs red; first to third segments with black markings; remaining rings with numerous irregular black lines; spiracles encircled by black, and with black streaks below them; hairs fine, whitish, not enlarged distally. On *Epilobium junceum*, *Oenothera*, *Balsaminum*. **Pupa** in cocoon of earth, in earth near the surface.

Distribution: New South Wales, Victoria, South Australia, Tasmania.

## *Phalaenoides thoracophora* TURNER

1920 *Phalaenoides thoracophora* TURNER, Trans. R. Soc. South Australia, Adelaide, **44**:125–126 (♀). – Terra typica: Claudie River, Queensland (Type, ♀, in Australian Museum, Sydney).

Diagnosis: Head yellow; apex of the frontal process and an interrupted bar between the antennae blackish; palpi yellow; upper and lower edge of 2nd joint of palpi blackish; 3rd joint blackish; antennae blackish; thorax yellow with black spots; abdomen brown, with the last urite and tuft yellow; legs blackish; tibiae and femora partly whitish beneath; hind tibiae and femora whitish. Fore wings blackish with bluish and purple reflections; a narrow whitish-grey subbasal fascia; median area whitish-grey, limited proximally by a line running from one sixth of costa to one quarter of dorsum, and distally by a line running from three fifths of costa nearly to tornus, containing a blackish line along the costa, a large oval antemedial spot, an incomplete fascia from beneath mid-costa, expanding on dorsum and continued to tornus; veins in distal area grey; cilia blackish. Hind wings blackish; a large basal white blotch not extending to the margins nor to the extreme base; cilia blackish, white at the extremities towards the tornus. Length of fore wing 22 mm.

Remark: Taxonomic position rather doubtful; may belong to the genus *Argyrolepidia*.

Distribution: North Queensland (Claudie River, III).

## Genus *Agarista* LEACH

(fig. 24)

1815 *Agarista* LEACH, Zool. Misc., **1**:30. – Species typica: *Agarista picta* LEACH = *Papilio agricola* DONOVAN (by monotypy).

Diagnosis: Proboscis fully developed; antennae moderately dilated towards extremity; palpi obliquely upturned to above vertex; 2nd joint fringed with long hair in front; 3rd slightly porrect; frons with conical truncate prominence, with raised rim at extremity; head, thorax, femora and abdomen with rough pilosity; tibiae with appressed hairscales; spurs long. Venation: in fore wing, vein 2 from about one-third of cell; 3, 4 and 5 well separated, equidistant; discocellulars incurved; 6 from upper cell-angle; areole small; 7 and 8 + 9 short-stalked, from extremity of areole; 10 apart, from upper angle of areole. Hind wings with vein 2 from 6/7 of cell; 3, 4 short-stalked or from a point; both branches of discocellulars incurved; 5 obsolescent, from middle of discocellulars; 6, 7 from a point.

Male genitalia: Uncus slightly longer than one-half of costa, very slender, faintly arched, blunt. Tubus analis narrowing distally, sclerified ventrally. Fold of vinculum narrow, fringed with longish, stiff hair. Valva elongate, narrow, arched, with costa and sacculus nearly parallel; apex rounded; fold of sacculus very broad, narrowing distally, reaching to just before two-thirds of valva, ending in a long, arched and hooked harpe reaching to or slightly beyond the apex of valva. Aedeagus somewhat shorter than costa, slender, rounded at the proximal extremity, nearly straight for

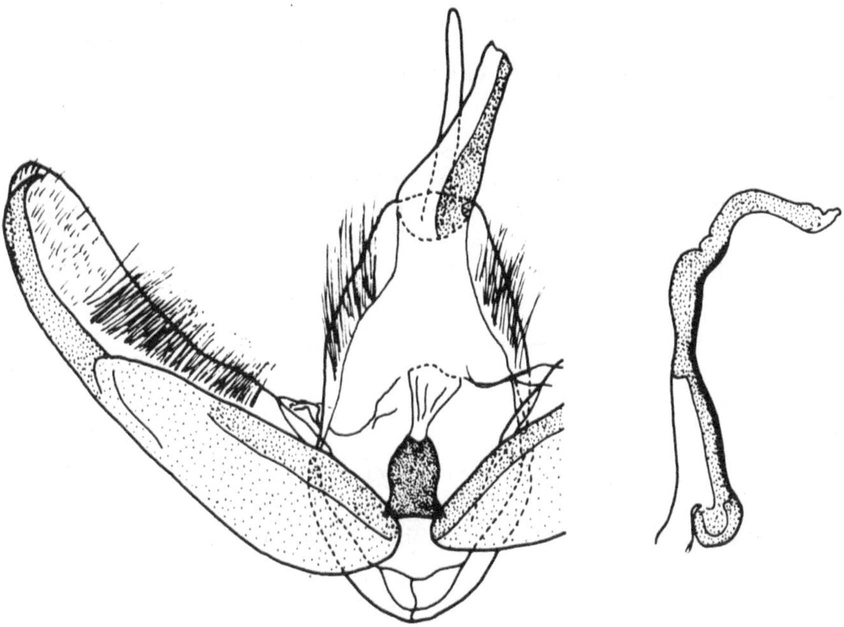

Fig. 24. *Agarista agricola* (DONOVAN). B. M. Noct. 8465

more than two-thirds, then curved; fultura inferior small, with the proximal margin straight and the sides incurved proximally, then outcurved; terminal margin excised at middle. Saccus short, rounded.

D i s t r i b u t i o n: Timor, New Guinea, Australia.

### Agarista agricola (LEACH)

1805 *Papilio agricola* DONOVAN, Insect. New Holland: pl. 32, fig. 1. – T e r r a t y p i c a: Australia (Type not designated).
1815 *Agarista picta* LEACH, Zool. Misc., 1:38; pl. 15. – T e r r a t y p i c a: Australia (Type not designated).
1854 *Agarista agricola* – WALKER, List Lepidopt. Het. Brit. Mus., 1:40.

D i a g n o s i s: Head and antennae black; palpi pale yellow with the third joint black; head marked with pale yellow on sides of frons and on vertex; thorax black in male, with a band on collar and a broad terminal band on tegulae, pale yellow; in female, whole thorax above pale yellowish with a bluish tinge; pectus and femora carmine red; fore and mid tibiae and tarsi spotted with white; mid tibiae streaked with orange, abdomen black with the anal tuft orange. Fore wings black; a broad basal streak of pale yellow reaching the inner line, in female very broad and occupying most of the basal area; a subbasal blue vertical line; inner line blue, excurved below costa; cell-extremity with a large oval orange spot; a postdiscal oblique series of orange spots; outer line formed of blue spots in costal half of the wing, and of narrow lunules running parallel to termen in dorsal half, with a spot just before tornus; a subterminal row of small spots; cilia tipped with white. Hind wings black with a postdiscal series of blue connected spots and typically a distal series of carmine spots from vein 4 to near tornus; cilia black at base, then white. Length of fore wing 30 – 35 mm.

Early stages: Larva white; head, appendages, a band between the third and the fourth rings and a large dorsal patch on the penultimate ring, orange red; the segments with a median black band; each segment with several black setae broadened and flattened towards the extremity. In last stages, the white colour becomes yellow, and ultimately dark orange. On *Vitis heterophylla*.

Distribution: Australia, New Guinea, Timor.

### Agarista agricola agricola (DONOVAN)

1912 *Agarista agricola agricola* – JORDAN in SEITZ, Großschmetterl. Erde, **11**:15; pl. 2e.

Diagnosis: As above, rather variable, female frequently with the spots reduced in number.

Distribution: New South Wales, Queensland.

### Agarista agricola biformis BUTLER

1884 *Agarista biformis* BUTLER, Ann. Mag. Nat. Hist., London, (5) **14**:404 (♂, ♀). – Terra typica: Port Darwin, North Australia (Types, ♂, ♀, in British Museum).
1901 *Agarista agricola biformis* – HAMPSON, Cat. Lepidopt. Phal. Brit. Mus., **3**:549.

Diagnosis: Fore wings with the orange spots paler in male, and becoming pale yellow in female; hind wings with the carmine band narrower and shorter, beginning at vein 3.

Distribution: North Australia, Nortwest Australia, Timor Laoet.

### Agarista agricola daemonis BUTLER

1876 *Agarista daemonis* BUTLER, Ann. Mag. Nat. Hist., London, (4) **17**:249 (♀). – Terra typica: Port Moresby, New Guinea (Type, ♀, in British Museum).
1901 *Agarista agricola daemonis* – HAMPSON, Cat. Lepidopt. Phal. Brit. Mus., **3**:550.

Diagnosis: Fore wings with the proximal yellow spots smaller in male; in female, the orange discal and outer spots obsolete; hind wings with the carmine band very reduced or absent.

Distribution: Papua, Northeastern New Guinea.

## A g a r i s t a   a g r i c o l a   t i m o r e n s i s   ROTHSCHILD

1896 *Agarista timorensis* ROTHSCHILD, Novit. Zool., 3:48 (♂). – T e r r a   t y p i c a: Oinainisa, New Guinea (Type, ♂, in British Museum).
1901 *Agarista agricola timorensis* – HAMPSON, Cat. Lepidopt. Phal. Brit. Mus., 3:550.

D i a g n o s i s: Fore wings with the basal pale streak broad in male; spot in cell reduced; inner and outer lines yellowish-white; in hind wings, the carmine band replaced by a series of blue greyish spots.

D i s t r i b u t i o n: Indonesian part of Timor.

## A g a r i s t a   a g r i c o l a   l a e t i o r   HULSTAERT

1924 *Agarista agricola* subsp. *laetior* HULSTAERT, Ann. Mag. Nat. Hist., London, (9) 13 (♂, ♀). – T e r r a   t y p i c a: Merauke, New Guinea (Types, ♂, ♀, in Leiden Museum).

D i a g n o s i s: In male, fore wings with the orange spots at base and in cell reduced; the postmedial series obsolete, more or less whitish; in hind wings, the blue band narrower than in *A. a. agricola*; the carmine area narrower and shorter. In female, fore wings with the space between the antemedian and postmedian lines below vein 2 to dorsum, blue and greenish; a triangular black patch on dorsum before middle; the whole basal part of the wing thus appearing broadly yellowish green and bluish.

D i s t r i b u t i o n: Southern Irian (Merauke = terra typica).

## Genus **Apina** WALKER

(fig. 25)

1855 *Apina* WALKER, List Lepidopt. Het. Brit. Mus., 3:756. – S p e c i e s   t y p i c a: *Apina callisto* WALKER (by monotypy).
1874 *Amazela* BOISDUVAL, Rev. Mag. Zool., (3) 2:66. – S p e c i e s   t y p i c a: *Amazela calisto* BOISDUVAL (by subsequent designation by HAMPSON, 1910).

D i a g n o s i s: Proboscis fully developed; antennae moderately pectinate in male, briefly bidentate and flattened ventrally in female; palpi porrect; second joint reaching nearly to middle of frons, fringed with hair in front; third short; frontal process long, naked, with acute angles, extremity truncate, incurved, frontal margin protruding, with a cutting edge; thorax clothed with hair and scales mixed; tibiae with long pilosity; hind tibiae with two pairs long spurs; abdomen smoothly scaled, without crests. Venation: in fore wings, vein 2 from two-thirds of cell; 3, 4, 5 separated, equidistant; 6 from lower cell-angle; upper discocellular faintly incurved; areole narrow; 7 and 8 + 9 from extremity of areole; 10 from upper angle of areole. Hind wings with vein 2 from three fifths of cell; 3, 4 and 6, 7 from angles of cell; discocellulars incurved; 5 obsolescent, from middle of discocellulars.

Male genitalia: Uncus relatively short, narrow, compressed, strongly bent near base, then nearly straight; terminal hook ill differentiated. Vinculum broad in proximal half, without bunches of stiff hair. Valva elongate, rather narrow; costa

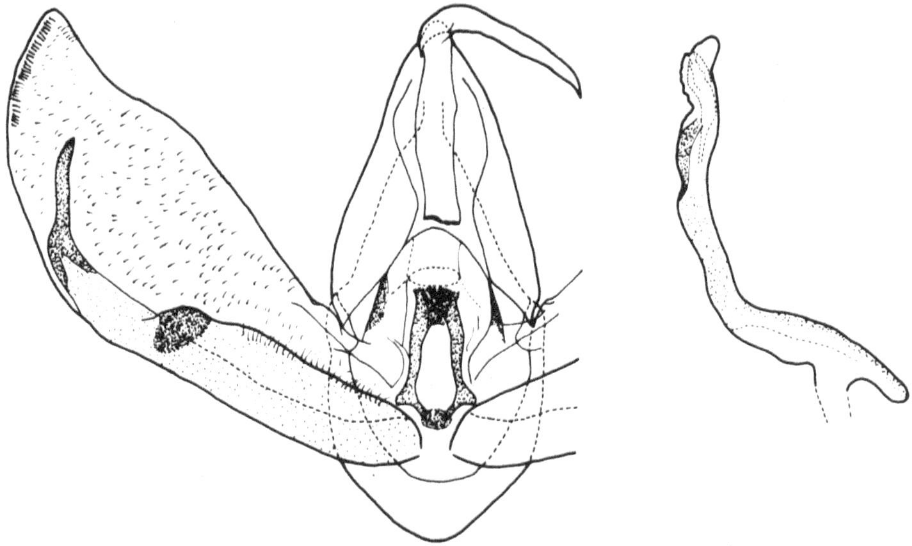

Fig. 25. *Apina callisto* WALKER. Type. B. M. Noct. 8296

nearly straight for about half of its lenght, then very faintly outcurved, but becoming incurved in terminal half; apex upturned, forming an angle of about 45°; termen and sacculus forming together a nearly regular curved line; a subapical series of stiff, curved setae; harpe forming a prolongation of the broad fold of sacculus, as long as one-third of valva, rather robust, slightly elbowed at middle, then narrowing. Aedeagus somewhat shorter than valva, slender for the distal two-thirds, bent at one-third and again at two-thirds, but then in an opposite direction, so that it forms an S; fultura inferior produced and narrowing distally, with the terminal margin very faintly outcurved, and bearing a row of minute spines; a patch of similar spines before the terminal margin. Saccus broadly rounded.

Distribution: Australia, Tasmania.

## Apina callisto WALKER

1855 *Apina callisto* WALKER, List Lepidopt. Het. Brit. Mus., 3:756 (♂). – Terra typica Tasmania (Type, ♂, in British Museum).
1874 *Amazela calisto* – BOISDUVAL, Rev. Mag. Zool., (3) 2:66.

Diagnosis: Head and thorax black marked with yellowish white and mixed with white hairscales; tegulae with orange hairscales at the extremity; fore femora with orange pilosity; tibiae and tarsi banded with white and black; abdomen orange; first urite black; edges of remaining urites black; underside white with black segmental edges. Fore wing reddish brown, sprinkled with yellowish white scales; lines yellowish; subbasal line short, reaching to about half the breadth of wing; inner line rather obsolescent, oblique, wavy, followed by a yellowish spot in cell; cell-extremity with a yellow spot edged with black; a yellow bar below it; outer line obsolescent, preceded by a large yellow spot beyond the discocellulars; subterminal line fine, parallel to termen; proximally, a series of rufous internervural spots; dorsal area tinged with rufous; a series of round yellowish subterminal spots. Hind wings blackish brown with a broad yellowish white discal band, excised proximally and faintly dentate distally; a subterminal series of yellowish spots. Length of fore wing 19 – 25 mm.

Distribution: Australia; Tasmania (Terra typica).

## Genus Eutrichopidia HAMPSON

(fig. 26)

1901 *Eutrichopidia* HAMPSON, Cat. Lepidopt. Phal. Brit. Mus., 3:554. – Species typica: *Phalaena latinus* DONOVAN (by original designation).

Diagnosis: Proboscis fully developed; antennae slightly dilated before the extremity; palpi with the second joint obliquely upturned and fringed with long hair in front; the third porrect, long, naked and dilated at the tip; frons with truncate conical prominence, with a raised rim at extremity; eyes hairy; head, thorax, femora, tibiae and abdomen with long pilosity; hind tibiae with a single

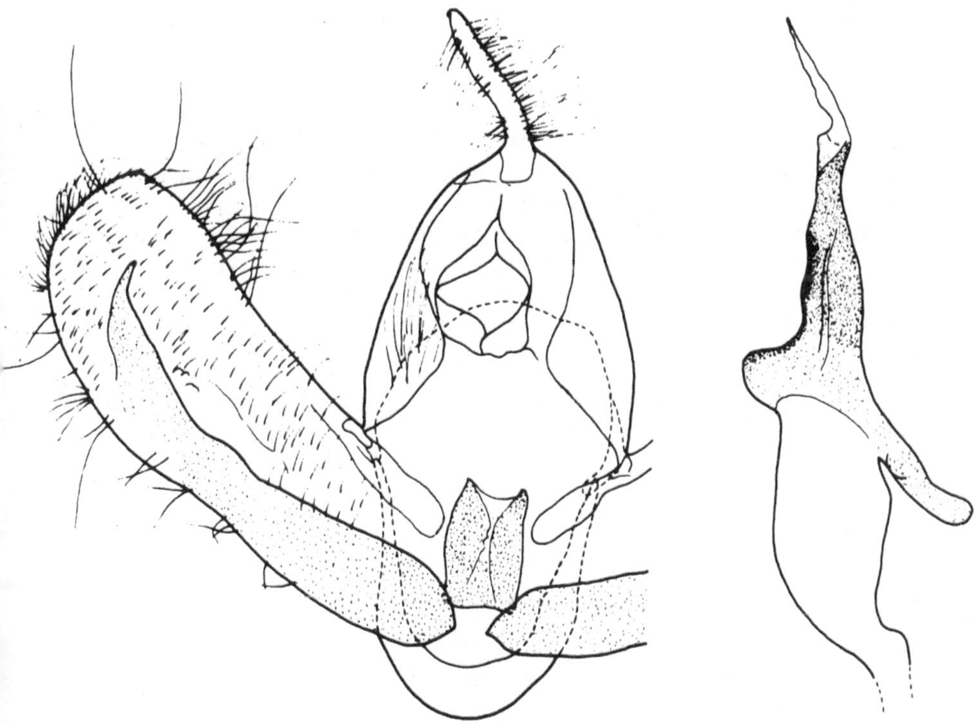

Fig. 26. *Eutrichopidia latina* (DONOVAN). B. M. Noct. 8378

pair spurs. Venation: in fore wing, vein 2 from two-thirds of cell; 3, 4 from angle of cell, separated from 5; areole elongate, with the sides parallel; 6 from areole, very near its base; 7 and 8 + 9 from extremity of areole; 10 from upper margin of areole near extremity. Hind wings with vein 2 from five eights of cell; 3, 4 and 6, 7 from angles of cell; discocellulars incurved; 5 obsolescent, from middle of discocellulars.

Male genitalia: Uncus relatively small, slender, faintly arched. Vincula narrow, with a few stiff hairs laterally. Valva elongate, lobe-shaped; costa faintly incurved; apex broadly rounded; sacculus faintly angled at two-thirds; fold of sacculus moderately broad, reaching to two-thirds of valva and ending in a small, hooked harpe; the very distal position of the harpe is characteristic and rather exceptional. Aedeagus slightly shorter than four fifths of valva, rather robust, but narrowing at both extremities, and appearing twisted at two thirds; fultura inferior with the sides parallel; proximal and distal margins slightly excised. Saccus short, semi-circular.

Distribution: Australia, Tasmania.

### **Eutrichopidia latina** (DONOVAN)

1805 *Phalaena latinus* DONOVAN, Insect. New Holland: pl. 32, fig. 4. – Terra typica: Australia (Type not designates).
1854 *Agarista latina* – WALKER, List Lepidopt. Het. Brit. Mus., 1:48.
1892 *Phalaenoides latina* – KIRBY, Syn. Cat. Lepidopt. Het.: 22.
1901 *Eutrichopidia latina* – HAMPSON, Cat. Lepidopt. Phal. Brit. Mus., 3:554; fig. 244 (♂, ♀).

Diagnosis: Body and wings bluish black; sides of palpi and of frons, basal joint of antennae, throat, spot on occiput, collar and tegulae pale yellow; coxae, fore femora and a tuft on the outer side of mid tibiae, with orange pilosity; tibiae and tarsi with white points; abdomen on the underside with the segments edged with white; anal tuft orange. Fore wings with the basal area with some bluish scales; an irregular ochreous band from just beyond middle of costa to tornus; veins bearing a number of bluish white scales in subterminal area; ciliae white, becoming yellowish apically and at tornus. Hind wings with a series of pale yellow terminal spots, interrupted on vein 5, with several black dots included therein; cilia pale yellow. Length of fore wing 23 mm.

Early stages: **Larva** pale yellow; head, appendages, spiracles and a large dorsal and lateral spot on the penultimate ring carmine red; below the spiracles runs a black line, emitting numerous lines which cover the sides and the dorsum of the segments. Pilosity fine, whitish. On *Haloragis teucrioides*. **Pupa** in a cocoon made of earth.

Distribution: Australia (New South Wales), Tasmania.

### Genus **Agaristodes** HAMPSON

(fig. 27)

1910 *Agaristodes* HAMPSON, Cat. Lepidopt. Phal. Brit. Mus., 9:456. – Species typica: *Agarista feisthameli* HERRICH–SCHÄFFER (by original designation).

D i a g n o s i s : Proboscis fully developed; antennae bipectinate with moderate branches, shorter towards the base and the apex in male, serrate in female; palpi porrect, extending to about the length of the head, and with long pilosity; frons with a rounded prominence, with a minute truncate process with raised rim at centre; thorax covered with rough hair, without crests; femora fringed with long hair; tibiae smoothly scaled; abdomen covered with rough hair, mostly with a basal crest. Venation: in fore wings, vein 2 from three quarters of cell; 3, 4 and 5 well separated, nearly equidistant; discocellulars well incurved; 6 from below the upper angle of cell; areole short; 7, 8 and 9 from tip of areole (8 and 9 not stalked); 10 from the upper edge of areole near its extremity. Hind wings with vein 2 from about five sevenths of cell; 3, 4 and 6, 7 from cell-angles; discocellulars well incurved; 5 obsolescent, from middle of discocellulars; 8 anastomosing with cell near base only. – Male genitalia: Uncus small, narrow, arched, with a terminal hook. Vincula expanded at middle, fused together for the most part of their length. Valva elongate, slightly broader distally; costa nearly straight; apex rounded; sacculus straight for more than half its length, then outcurved; fold of sacculus rather broad, reaching for two

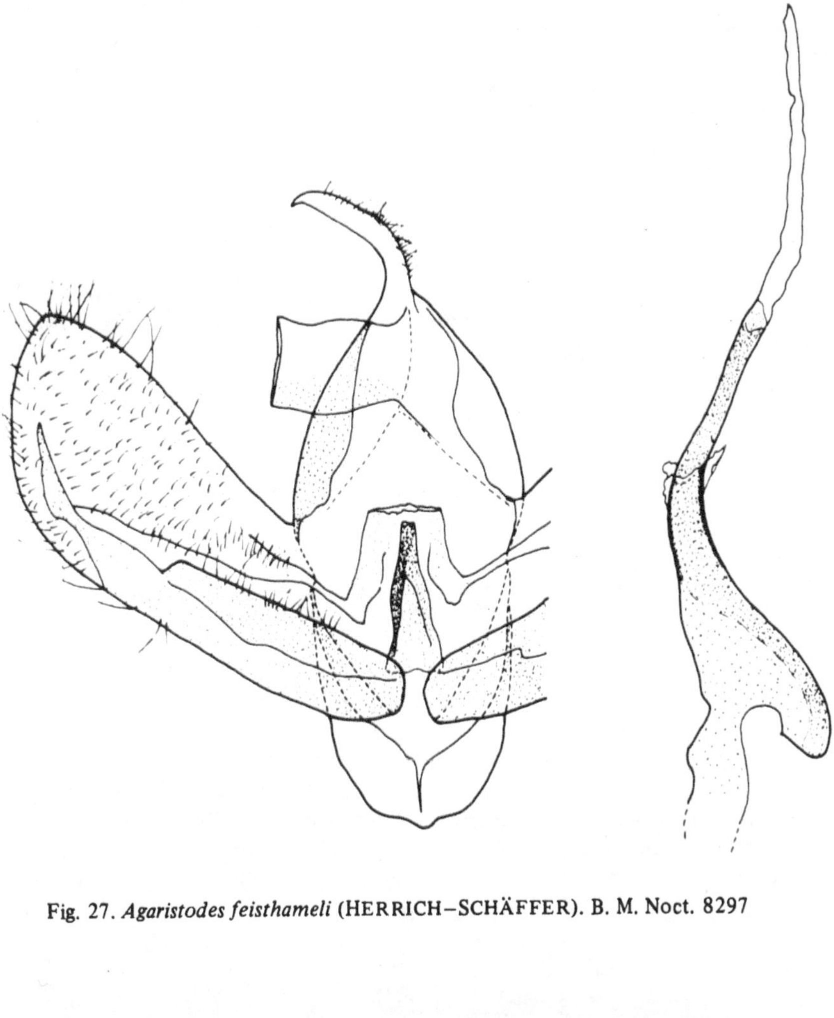

Fig. 27. *Agaristodes feisthameli* (HERRICH–SCHÄFFER). B. M. Noct. 8297

thirds of valva, ending in a slender, very faintly arched harpe, slightly expanded after the base. Aedeagus nearly as long as the valva, rather slender, especially distally, straight in basal half, then bent at an angle of about 120°, and again straight; fultura inferior narrowly triangular, with the top cut off. Sacculus nearly semi-circular, slightly produced at middle.

Distribution: Tasmania.

### **Agaristodes feisthameli** (HERRICH–SCHÄFFER)

1853 *Agarista feisthameli* HERRICH–SSCHÄFFER, Außereurop. Schmetterl.; fig. 25. – Terra typica: Tasmania (Type, ♂, in British Museum).
1910 *Agaristodes feisthameli* – HAMPSON, Cat. Lepidopt. Phal. Brit. Mus. 9:456; fig. 219 (♂, ♀).

Diagnosis: Head and appendages, and thorax black; palpi marked with white in female; collar and tegulae streaked with white, broader in female; femora with orange pilosity; legs varied with black and white; abdomen black, with basal and anal crests, and edges of two last urites orange. Fore wings blackish brown, with bluish white reflections; lines bluish white: subbasal line wavy; inner line oblique, wavy; beyond it, a yellowish white spot in cell; a very large yellowish white patch from below two-thirds of costa, occupying the distal half of cell and the bases of interspace II–VI, interrupted in its fore part by a patch of ground colour on the discocellulars; in female, the large patch in separated in two by brown on the discocellulars; outer and subterminal bluish white lines fine, only reaching vein 6; a terminal series of white dots; cilia tipped with white in apical and tornal ares. Hind wings blackish brown; a large yellowish white area occupying the middle of wing, narrowing towards the anal angle, reduced in female; before end of cell, a narrow white line; cilia with white tips in apical and tornal areas. Length of fore wings 18 – 22 mm.

Distribution: Tasmania.

### Genus **Periscepta** TURNER

(fig. 28)

1920 *Periscepta* TURNER, Trans. R. Soc. South Australia, Adelaide, **44**:126. – Species typica: *Agarista polysticta* BUTLER (by original designation).

Diagnosis: Proboscis fully developed; antennae filiform, reaching to middle of costa; frons with a conical prominence with a raised rim at the apex; palpi short, porrect; 2nd joint hairy; 3rd very short, hairy; hind tibiae hairy above; abdomen hairy, without crests. Venation: in fore wings, vein 2 rather near the cell-angle; 3, 4, 5 well separated, equidistant; discocellulars faintly incurved; 6 from upper angle of cell; areole narrow; 7 and 8 + 9 from a point at the extremity of areole; 10 from the upper margin of areole, near its extremity. In hind wings, vein 2 from rather near the cell-angle; 3, 4 and 6, 7 from the angles of cell; discocellulars incurved; 5 obsolescent, from middle of discocellulars.

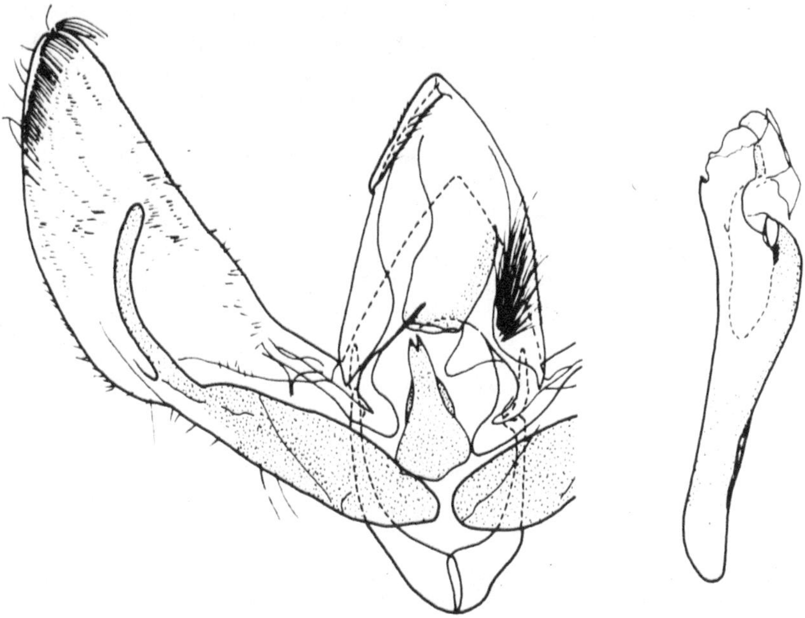

Fig. 28. *Periscepta polysticta* (BUTLER). B. M. Noct. 8440

Male genitalia: Uncus relatively small, narrow, compressed, nearly straight, ending in a weak hook; length about one quarter of the costa. Tubus analis short and broad, sclerified ventrally. Vincula moderately broad, with a bunch of stiff hair near the proximal end. Valva elongate, rather broad, especially in the median portion; costa faintly incurved beyond the base, then nearly straight; apex blunt; sacculus very faintly incurved at the base, becoming straight to end of fold, then broadly outcurved; apex and about the terminal fifth of sacculus with a row of stiff marginal spinules; fold of sacculus rather broad basally, slightly narrowing distally to reach about mid-valva, ending in a narrow, arched, blunt harpe about one-third of costa long. Aedeagus about as long as the costa, rather slender proximally, but broadening distally; terminal portion rather irregularly shaped; fultura inferior roughly triangular, with the proximal margin faintly produced at middle; sides faintly incurved; distal edge bifid. Saccus short, rounded.

D i s t r i b u t i o n: Australia.

## Periscepta polysticta (BUTLER)

1875 *Agarista polysticta* BUTLER, Ann. Mag. Nat. Hist., London, (4) **15**:138 (♂, ♀). – T e r r a t y p i c a: Sydney, New South Wales (Type, ♀, in British Museum).
1892 *Phalaenoides butleri* SWINHOE, Cat. Het. Mus. Oxon.:157; pl. 5, fig. 6 (♂). – T e r r a t y p i c a: Australia (Type, ♂, in Museum Oxford University).
1892 *Phalaenoides polystictus* – KIRBY, Syn. Cat. Lepidopt. Het.:22.
1901 *Phalaenoides polysticta* – HAMPSON, Cat. Lepidopt. Phal. Brit. Mus., **3**:560.
1920 *Periscepta polysticta* – TURNER, Trans. R. Soc. South Australia, Adelaide, **44**:126.

D i a g n o s i s : Body and wings black; 1st and 2nd joints of palpi, sides of frons, 1st antennal joint, occiput, collar, tegulae and thorax marked with pale yellow; throat yellow; fore coxae and mid tibiae with a tuft of orange hair; terminal urites, anal tuft and ventral bands on four last urites orange. Fore wings with two yellow basal dots; a subbasal and two antemedial series of three spots; a medial spot below costa, one in cell-end, and one below it on vein 1; beyond the cell, a band from below costa to vein 2, excurved below costa and more strongly beyond the lower cell-angle; a small spot below it on vein 1, and usually three spots beyond it below costa; a subterminal series of dots; cilia white apically and at tornus. Hind wings unspotted, or with a pale yellow band beyond the cell between veins 2 and 6; cilia pale yellow, except between veins 4 and 6. Length of fore wing 18 – 20 mm.

D i s t r i b u t i o n : New South Wales, West Australia.

## Genus **Pimprana** Moore

(fig. 29)

1879 *Pimprana* MOORE, Descr. lepidopt. Ins. ATKINSON:41. – S p e c i e s t y p i c a : *Pimprana atkinsoni* MOORE (by monotypy).

D i a g n o s i s : Proboscis fully developed; antennae setose, faintly dilated distally in second half, with rather longisch ciliae and setae; palpi stout, densely pilose, porrect, extending slightly beyond the head; 3rd joint short; eyes hairy; frons with

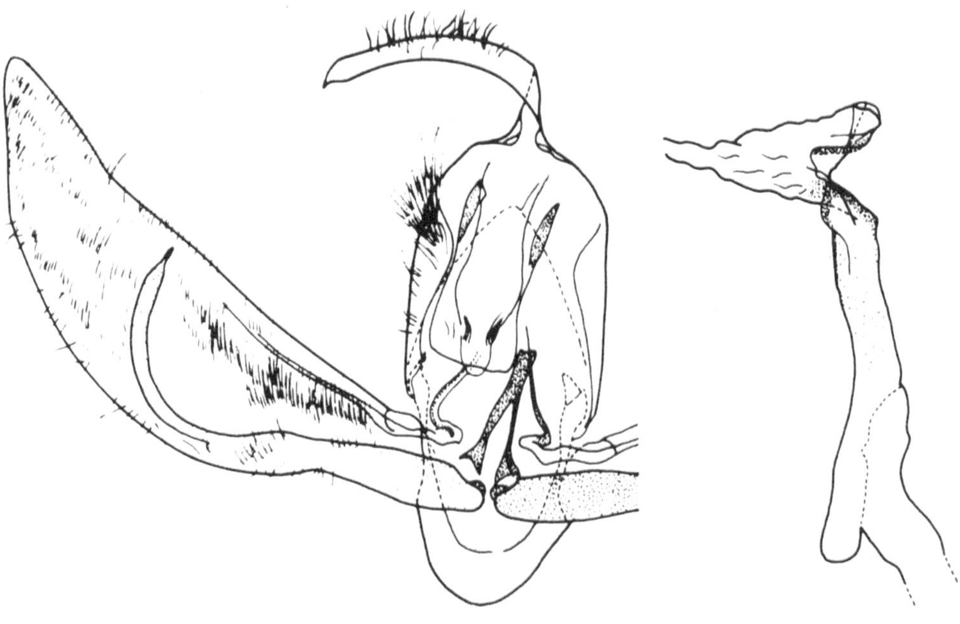

Fig. 29. *Pimprana atkinsoni* MOORE. B. M. Noct. 8394

truncate conical prominence with a raised rim at the extremity; thorax above and below, femora and tibiae with long hair; abdomen without distinct hair-tufts above; sides with hair-tufts; in female, tip of abdomen thickened and woolly. Venation: in fore wings, vein 2 from the four fifths of the cell; 3, 4, 5 much approximated; discocellulars incurved; 6 from the middle of areole; 7 and 8 + 9 from tip of areole; 10 from upper side of areole. Hind wings with vein 2 near cell-angle; 3, 4 and 6, 7 from the cell-angles; 5 obsolescent, from middle of discocellulars.

Male genitalia: Uncus elongate, slightly longer than half of costa, narrow, compressed, arched, with a minute terminal hook. Tubus analis broad, with lateral sclerifications at the base. Vincula rather narrow, with a narrow fold, covered with rather soft hair. Valva elongate, banana-shaped; apex semi-acute; sacculus more or less straight for nearly one-quarter, then arched; fold of sacculus narrow, reaching to two-fifths of sacculus, ending in a slender, well-arched harpe, slightly shorter than the uncus, with an acute terminal point. Aedeagus about two-thirds of costa in length, rather slender, very faintly bent at middle; terminal portion and base of vesica bearing agglomerations of microscopic cornuti; fultura inferior narrow, produced distally, with the narrow terminal portion cut nearly straight. Saccus short semi-elliptical.

Distribution: India (Sikkim).

## **Pimprana atkinsoni** MOORE

1879 *Pimprana atkinsoni* MOORE, Descr. lepidopt. Ins. ATKINSON:42; pl. – Terra typica: Darjeeling, Sikkim (Type, ♂, in British Museum).

Diagnosis: Palpi orange with a black lateral blotch; head and thorax above orange olive; underside and pilosity of legs orange; tarsi pale yellowish; abdomen black above with some orange hair at base; sides, underside and anal tuft orange. Fore wings orange olive; basal area more or less strongly sprinkled with whitish; a pale yellowish basal spot on dorsum, and a similar costal mark, somewhat less distinct; basal area limited by the inner line which is only visible from dorsum to cell, arched, deep Indian red with a paler medial line; marks in cell and on discocellulars very obsolete, partly edged with silvery white; edge of the discocellular marks sending a streak towards the termen; lower vein of the cell and bases of veins 2, 3 and 4 also silvery white; outer line similar to the inner one, only visible from dorsum to vein 4, incurved, running from three quarters of dorsum to termen just below vein 4; a darker red triangular spot below apex; a series of silvery subterminal streaklets on red ground, in interspaces II, III, VI and VII, and a silvery broken terminal line; cilia concolorous. Underside yellow orange except in inner area, with two cell spots and the discocellular mark. Hind wings blackish chocolate, with an indistinct darker discocellular spot; orange terminal band, 1 – 2 mm broad, occupying the costa, and slightly produced inwards in interspace II. Underside the orange band broader, and the discocellulars with an elongate squarish spot. Length of fore wing 15 mm. Female: the Indian red lines less distinct, otherwise like the male, but with a large, woolly anal tuft.

Distribution: Sikkim (Darjeeling = terra typica).

## Genus **Syfania** OBERTHÜR

(fig. 30)

1893 *Syfania* OBERTHÜR, Etud. Entomol., **18**:19. – Species typica: *Agarista bieti* OBERTHÜR (by subsequent designation: HAMPSON 1901).

Diagnosis: Proboscis fully developed; antennae dilated before extremity; palpi obliquely upturned, fringed with long hair below; frons with a truncate conical prominence with raised rim at the extremity; pectus and underside of abdomen with long pilosity; tibiae with the spurs long; hind tibiae with long hair above and with a few spines between the medial and terminal spurs. Venation: in fore wings, vein 2 from about three-quarters of cell; 3, 4, 5 separated and equidistant; areole short; 6 from areole; 7 and the very short stalk of 8 + 9 from extremity of areole; 10 from the upper angle of areole. Hind wings with vein 2 from four-sevenths of cell; 3, 4 and 6, 7 from the cell-angles; discocellulars angled inwards at middle; 5 obsolescent, from middle of discocellulars.

Male genitalia: Uncus about seven-tenths of costa in length, arched basally, very slender and nearly straight distally. Tubus analis narrow, sclerified both dorsally and ventrally. Fold of vinculum rather broad and angled at middle, with long stiff

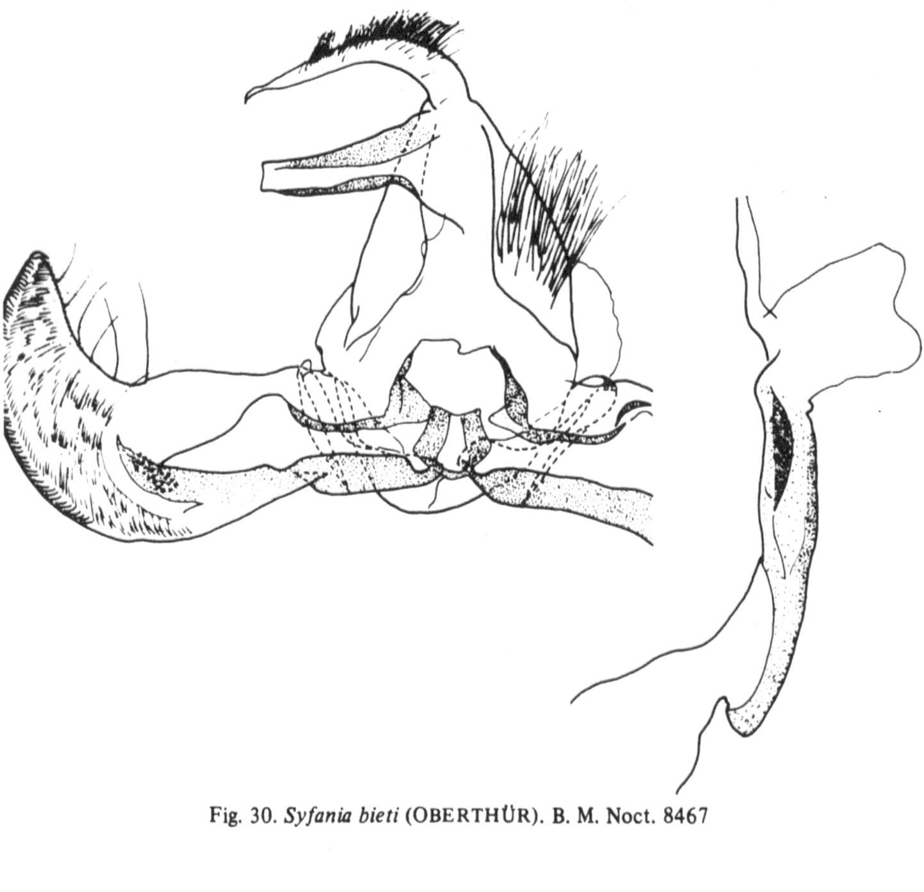

Fig. 30. *Syfania bieti* (OBERTHÜR). B. M. Noct. 8467

hair. Valva broad, upcurved; apex pointed; margin of sacculus edged for the distal half with short stiff spinules; fold of sacculus very narrow, reaching to slightly beyond one-third, and ending in a short, broadish, slightly arched harpe covered with sparse denticuli; base of valva bearing a club-shaped structure. Aedeagus distinctly longer than costa, rather slender, hardly arched, broadened terminally; a subterminal patch of denticuli; fultura inferior short, escutcheon-shaped; fultura superior with narrow lateral processes connected to a fold on the valvula. Saccus hardly developed, rounded.

Distribution: China.

Key to the species

1 Hind wings with a yellow or orange bar at tornus . . . . . . . . . . . . . . . . . . . . . . 2
1' Hind wings without yellow or orange bar at tornus . . . . . . . . . . . . . . . . . . . . . . 3
2 Hind wings yellow, with a black discocellular spot, a fascia on anal area and terminal band . . . . . . . . . . . . . . . . . . . . . . . . . . . . . . . . . . . **S. dubernardi,** p. 135
2' Hind wings black, with yellowish spots in, below and beyond the cell . . . **S. dejeani,** p. 136
3 Abdomen with yellow segmental bands or lines . . . . . . . . . . . . . . . . . . . . . . 4
3' Abdomen without yellow abdominal lines . . . . . . . . . . . . . . . . . . **S. bieti,** p. 135
4 Hind wings with spots yellowish, separate . . . . . . . . . . . . . . . . **S. giraudeaui,** p. 136
4' Hind wings with spots white, conjoined to form a broad medial area . **S. oberthueri,** p. 137

## Syfania bieti (OBERTHÜR)

1886 *Agarista bieti* OBERTHÜR, Bull. Soc. Entomol. France, Paris, (6) **5**:227. – Terra typica: Tatsienlou, Western China (Type, ♂, in British Museum).
1892 *Androloma bieti* – KIRBY, Syn. Cat. Lepidopt. Het.:35.
1901 *Syfania bieti* – HAMPSON, Cat. Lepidopt. Phal. Brit. Mus., **3**:633; fig. 278 (♂, ♀).

Diagnosis: Body and wings black; palpi banded with yellow; sides of frons yellow; collar, tegulae and thorax above streaked with yellow; legs with orange pilosity; underside of abdomen orange. Fore wings with a leaden streak on basal half of vein 1; a leaden spot in middle of cell; a leaden discocellular bar; an orange spot in middle, in and below cell-end; a postmedial band from below costa to vein 3; streaks of orange scales below cell and on dorsum; two spots below base of costa; patches round the leaden marks; a double finely undulate postmedial line, strongly dentate inwards on vein 2; pairs of small wedge-shaped subterminal spots on each side of the veins. Hind wings with white spots in middle and below cell-end; a short bar just beyond the discocellulars; cilia white at the extremities. Length of fore wing 16 – 17 mm.

Distribution: China (Central Szechuan [Tatsienlou = terra typica], Northern Szechuan, Kansu).

## Syfania dubernardi OBERTHÜR

1894 *Syfania dubernardi* OBERTHÜR, Etud. Entomol., **18**:21; pl. 8, fig. 70 (♀). – Terra typica: Tchang-Kou, Western China (Type, ♀, in British Museum).
1936 *Syfania dubernardi taipeishanis* MELL, Stettin. Entomol. Ztg., **97**:85. – Terra typica: Tapaishan, China (Type in Berlin Museum).
1943 *Syfania dubernardi kansuina* BRYK, Entomol. Tijdschr., **68**:150. – Terra typica: Kansu, China.

D i a g n o s i s : Body and fore wings black; collar and tegulae streaked with orange; abdomen with an orange dorsal streak on the first two urites; underside with pale yellow pilosity. Fore wings with the markings bluish white in male, pale yellow in female, viz.: a spot at base of cell; a streak below cell; a spot in cell-end; an elongate spot below cell; streak on basal half of vein 1; a postmedial band from below costa to vein 3; a small spot above tornus. Hind wings bluish white in male, orange yellow in female; base and fascia above anal margin black; a large black discocellular spot; a broad terminal black band enclosing a pale bar from the submedian fold to tornus. Length of fore wing 18 mm.

D i s t r i b u t i o n : China (Tsang Ku = terra typica; Eastern Szechuan? Northwestern Yunnan?; Southern Shensi, Kansu).

## Syfania dejeani (OBERTHÜR)

1893 *Syfania dejeani* OBERTHÜR, Etud. Entomol., **18**:19; pl. 5, fig. 68 (♂). – T e r r a t y p i c a : Tatsienlou, West China (Type, ♂, in British Museum).

D i a g n o s i s : Body and wings black; palpi banded with yellow; sides of frons yellow; collar, tegulae, thorax above striped with yellow; pectus and legs with orange pilosity; abdomen with a fine dorsal yellowish streak; underside with orange pilosity. Fore wings with some metallic grey, viz. a streak on basal half of vein 1; a spot in middle of cell, and a discocellular bar; orange yellow spots in middle and in and below cell-end; a postmedial band from below costa to vein 3; streaks of yellow scales below cell and on dorsum; spots below base of costa and some scales round the metallic grey marks; postmedial line double, strongly dentate inwards in vein 2; pairs of small wedge-shaped orange subterminal marks on each side of the veins. Hind wings with yellow spots in middle and below end of cell; a bar just beyond the discocellulars; an orange bar from the submedian fold to tornus; anal margin fringed with orange; cilia yellow from apex to vein 2. Length of fore wing 16 – 18 mm.

D i s t r i b u t i o n : Szechuan (Tatsienlou = terra typica), Northwest Yunnan.

## Syfania giraudeaui OBERTHÜR

1893 *Syfania giraudeaui* OBERTHÜR, Etud. Entomol. **18**:19; pl. 5, fig. 74. – T e r r a t y p i c a : Tatsienlou, West China (Type in British Museum).

D i a g n o s i s : Body and wings black; palpi banded with yellowish white; sides of frons pale yellow; collar, tegulae and thorax above streaked with pale yellow; pectus and femora with pale yellow pilosity; tibiae with orange pilosity; abdomen with pale yellow dorsal segmental lines; underside orange. Fore wings with some leaden marks, viz.: a streak on basal half of vein 1; a spot in middle of cell; and a discocellular bar; orange spots below base of costa, in middle and below end of cell; streaks below cell and on base of the dorsal area; a postmedial band from below costa to vein 3; some yellow scales round the leaden marks; a double postmedial line, strongly angled inwards on vein 2; pairs of small wedge-shaped subterminal spots on each side of the veins; a more prominent spot above tornus; cilia white apically. Hind wings with a pale yellow spot in cell-middle; a square spot below the cell-extremity; a large patch beyond cell; a subterminal spot below vein 2; and a series of points between it and the apex. Length of fore wing 15 – 16 mm.

Distribution: Szechuan (Tatsienlou = Terra typica).

### Syfania oberthueri ALPHERAKY

1895 *Syfania oberthüri* ALPHERAKY, Dtsch. entomol. Z. Iris, Dresden, **8**:184 (♂, ♀). – Terra typica: Kham Country, Northwest China (Types, ♂, ♀, in Leningrad Museum?).
1901 *Syfania giraudeaui* aberr. 1 – HAMPSON, Cat. Lepidopt. Phal. Brit. Mus., **3**:633.

Diagnosis: Very near *S. dejeani* OBERTHÜR, but differing in the pattern of the hind wings. Ground colour yellowish white; the small spot in the black marginal band, near the anal angle, together with the anal margin, yellowish white, not orange; the mark on the discocellulars well separated from the black margin in male, but narrowly fused with the black margin in female, leaving large spots of ground colour.

Distribution: China (Kham country = terra typica).

## Genus Aegocera LATREILLE

(fig. 31)

1809 *Aegocera* LATREILLE, Gén. Crustac. et Insect., **4**:211. – Species typica: *Noctua venulia* CRAMER (by monotypy).

Diagnosis: Proboscis fully developed; antennae strongly dilated between middle and extremity; palpi upturned with the second joint bearing in front long pilosity, produced at the extremity to an acute angle; third joint long, naked, somewhat porrect; eyes naked; frons with a truncate conical projection with a raised rim at the extremity; tibiae fringed with long rough hair above; abdomen with basal crests on the first urite; male with lateral tufts of long hair on the last urite, and protrusible tufts from base. Fore wings of male with more or less developed subcostal groove of ribbed hyaline membrane; a clicking sound being produced during flight, probably by friction with the tarsal spines. Venation: in fore wings, vein 2 from three-quarters to four-fifths of cell; 3, 4, 5 approximated, equidistant or vein 5 slightly more apart; discocellulars slightly incurved; 6 from cell-angle; areole short; 7 and the stalk of 8 + 9 from tip of areole; 10 from upperside of areole, near the extremity. In hind wings, vein 2 from two-thirds to four-fifths of cell; 3, 4 and 6, 7 from the cell-angles; discocellulars faintly incurved; 5 obsolescent, from middle of discocellulars.

Male genitalia: Uncus small, slightly shorter than half of costa, hooked in terminal third, with short stiff pilosity. Tubus analis short, sclerified. Vincula rather broad, fringed in middle portion of the fold by rather long, stiff hair. Valva narrow,

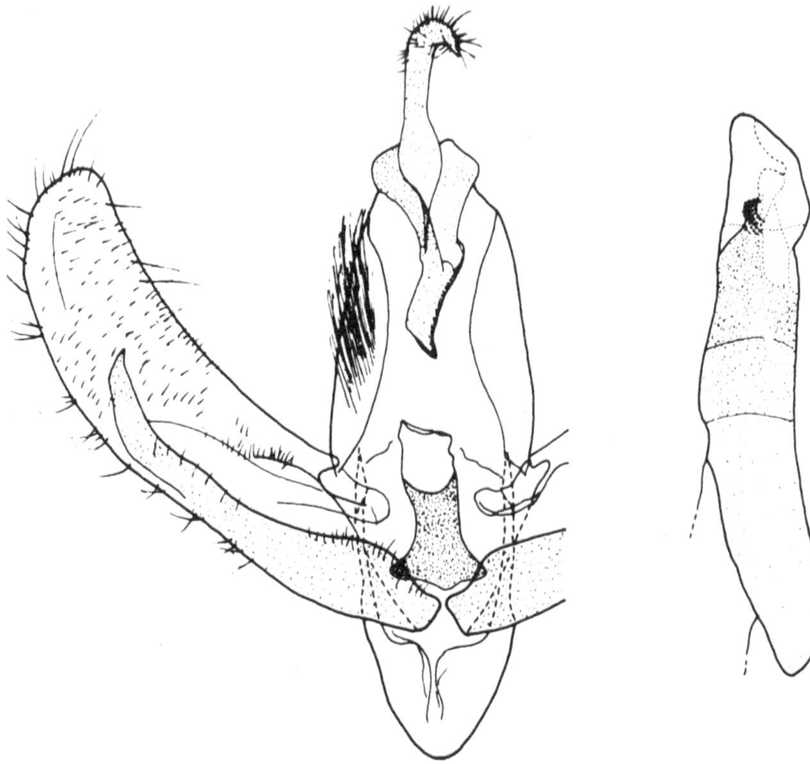

Fig. 31. *Aegocera venulia* (CRAMER). B. M. Noct. 8285

banana-shaped, with costa and sacculus parallel, and the apex rounded; fold of sacculus narrow, reaching to three-fifths of valva and ending in a short, faintly arched harpe. Aedeagus very slightly longer than valva, robust, faintly bent at middle; terminal lobe bearing a small patch of minute cornuti; fultura inferior broad proximally, but soon narrowing, with the sides nearly parallel. Saccus semi-elliptical.

Distribution: India, Burma, Ceylon.

Key to the species

1 Fore wings deep red, sprinkled with yellowish; a pale basal fascia and two oblique spots, one in cell-end and one beyond cell . . . . . . . . . . . . . . . . . . . . . . ***A. tripartita***, p. 140

1′ Fore wings with the ground colour deep reddish brown or dark brown, sprinkled with yellowish . . . . . . . . . . . . . . . . . . . . . . . . . . . . . . . . . . . . . . . . . . 2

2 Fore wings with a pale fascia from base to well before termen, broadened at middle of wing, with a black ocellate spot in broad section of fascia . . . . . . . . . ***A. venulia***, p. 139

2′ Fore wings with a pale T-shaped fascia in basal half, and an oblique patch beyond it ***A. bimacula***, p. 139

### Aegocera venulia (CRAMER)

1777 *Noctua venulia* CRAMER, Pap. Exot., **2**:pl. 165 D. – Terra typica: India (Type not designated).
1809 *Aegocera venulia* – LATREILLE, Gén. Crustac. et Insect., **4**:212.

Diagnosis: Head white; second joint of palpi spotted with brown and black; third joint black at the extremity; antennae and sides of frons black; thorax white, striped with brown on collar; edges of tegulae and thorax with subdorsal blackish and orange stripes; tibiae and tarsi orange, banded with black and white; abdomen orange with a dorsal series of black spots, and the underside whitish. Fore wings dark brown, thickly sprinkled with ochreous; a creamy white fascia from base through cell to near termen in interspace V, broadening below middle of cell to reach vein 1, with an ocellate black spot with the centre silvery blue, on it, its upper edge indented by the rusty-red cell- and discocellular spots, which are also centered with silvery blue; a short rusty-red streak ending in silvery blue scaling above base of vein 1; terminal part of fascia edged with black and rusty-red; terminal line rusty-red; cilia black, tipped with white. Hind wings orange; a rusty-red discocellular spot and terminal band, broader towards apex; cilia whitish with a dark brown median line. Length of fore wing 15 – 17 mm.

Distribution: Southwest Pakistan, Northwest India to Burma and Ceylon.

### Aegocera bimacula WALKER

1854 *Aegocera bimacula* WALKER, List Lepidopt. Het. Brit. Mus., **1**:57 (♀). – Terra typica: India (Type, ♀, in British Museum).

Diagnosis: Head and thorax white; antennae, sides of the second and extremity of the last palpal joints, and sides of front black; collar striped with black; edges of tegulae and stripes on thorax black; pectus and legs orange; fore and mid tibiae, and tarsi with black spots in front; abdomen orange with dorsal series of black spots, and sublateral series of dots; anal hair deep chocolate brown. Fore wings deep reddish brown sprinkled with ochreous; markings placed on a deep chocolate-red patch, extending to beyond middle of vein 1 and beyond cell; a yellowish spot at base of costa, mostly connected with an antemedial spot in and below cell, and forming with it a wedge-shaped patch, often connected by a streak with an oblique bar across cell-end, ending in the submedian fold; an oblique lunulate band beyond cell between veins 3 and 8; all these markings edged with black; some silvery blue markings, viz., in and below middle of cell; a discocellular lunule; and a few scales at the extremity of the deep chocolate-red; cilia dark brown. Hind wings orange; markings deep chocolate-red in male, blackish brown in female, viz., spots at cell-angles; a terminal band, strongly excised at middle, broader towards apex and tornus, with an orange spot on termen in interspace I; sometimes a spot at the lower cell-angle, connected with the terminal band; cilia orange. Length of fore wing 21 – 22 mm.

Distribution: Southwest Pakistan, Nortwest India to Burma and Ceylon.

### Aegocera tripartita KIRBY

1880 *Aegocera tripartita* KIRBY, Proc. R. Dublin Soc., (2) **2**:340. – Terra typica: Burma (Type, ♂, in British Museum).

Diagnosis: Head and thorax white; antennae and palpi except in front, forepart of frons, and some scales on vertex black; collar striped with black; edges of tegulae and subdorsal stripes on thorax black and orange; tibiae orange, circled with black; tarsi black, ringed with white; abdomen orange, with dorsal series of black spots; underside whitish. Fore wings deep red, strongly sprinkled with ochreous; a wedge-shaped yellowish white fascia on the media from base to middle of cell; a roughly elliptical spot in cell, extending to just below it; a triangular patch beyond cell; all these markings separated by clear red patches with silvery blue bars on them; the median bar joining a silvery spot on a triangular red patch below cell; cilia whitish at tips. Hind wings orange; markings red-brown in male, black in female; a broad costal stripe, conjoined to a discocellular lunule; a broad terminal band, slightly narrower at middle; anal area brown; cilia whitish. Length of fore wing 15 – 16 mm.

In this species, the subcostal ribbed membrane is particularly well developed, and the neighbouring veins are strongly downcurved. The male genitalia, however, do not show any particularities.

Distribution: Burma.

## Genus Platagarista JORDAN

(fig. 32)

1912 *Platagarista* JORDAN in SEITZ, Großschmetterl. Erde, **11**:21. – Species typica: *Agarista tetrapleura* MEYRICK (by original designation).

Diagnosis: Proboscis fully developed; antennae faintly expanded between middle and apex, and depressed ventrally; eyes naked; 1st and 2nd joints of palpi fringed with rough hair, 3rd elongate in both sexes, naked; frons with truncate conical prominence with a raised rim at the extremity; tibiae with appressed scales; 1st joint of hind tarsi elongate, curved and dilated, bearing on outer side a rim with minute serrations. Wings in male with streaks of ribbed semi-hyaline membrane in and below cell in fore wings, and below vein 8 and in cell in hind wings. Venation: in fore wings, vein 2 from about three-quarters of cell; 3, 4, 5 very much approximated; discocellulars strongly incurved; 6 from the upper cell-angle; areole narrow; 7 and stalk of 8 + 9 from the tip of areole; 10 from upper side of areole near its extremity. In hind wings, vein 2 from two-thirds of cell; 3, 4 and 6, 7 from the cell-angles; discocellulars faintly incurved; 5 obsolescent, from middle of discocellulars.

In flight, the modified tarsi serve to produce a sound through pression against the transversely ribbed hyaline spots.

Male genitalia: Uncus relatively very small and short, depressed, ending in an acute point and bearing sparse hair. Tubus analis very short. Vinculum rather narrow, with the fold triangular and covered with loose pilosity. Valva rather broad; costa incurved, more or less parallel with sacculus; apex semi-acute; fold of sacculus rather

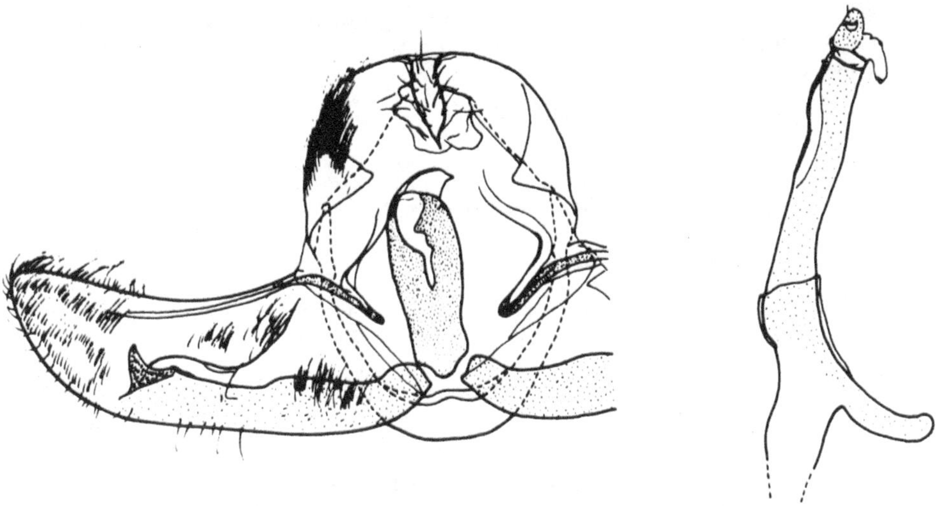

Fig. 32. *Platagarista tetrapleura* (MEYRICK). B. M. Noct. 8447

narrow, reaching to two-thirds of valva and bearing terminally an obliquely placed small triangular harpe. Aedeagus about as long as the valva, rather slender, arched, bearing terminally a lateral series of hardly visible cornuti; fultura inferior narrow, elongate-elliptical. Saccus hardly differentiated, broadly rounded.

D i s t r i b u t i o n: Australia.

## Platagarista tetrapleura (MEYRICK)

1891 *Agarista tetrapleura* MEYRICK, Trans. R. Soc. South Australia, Adelaide, **14**:195 (♀). – T e r r a t y p i c a: Australia (Type, ♀, in British Museum).

1891 *Phalaenoides albamedia* LUCAS, Proc. Linn. Soc. New South Wales, Sydney, (2) **6**:301. – T e r r a t y p i c a: New South Wales (Type, ♂, in South Australian Museum, Adelaide).

1892 *Phalaenoides macleayi* SWINHOE, Cat. Het. Mus. Oxon.:157; pl. 5, fig. 11 (♀). – T e r r a t y p i c a: North Australia (Type, ♀, in Museum Oxford University).

1892 *Phalaenoides tetrapleura* – KIRBY, Syn. Cat. Lepidopt. Het.:897.

1912 *Platagarista tetrapleura* – JORDAN in SEITZ, Großschmetterl. Erde, **11**:21; pl. 2g.

D i a g n o s i s: Body and wings black; basal joint of palpi, sides of frons, postocular region marked with pale yellow; 2nd joint of palpi streaked with white; 1st joint of antennae, occiput, collar, tegulae and thorax above with pale yellow points; pectus with orange pilosity; tibiae and tarsi with white points; abdomen with yellowish-white segmental rings; anal tuft and basal lateral tufts orange. Fore wings sprinkled with a few-bluish white scales; an oblique series of six small yellow spots, somewhat irregularly placed, from costa slightly beyond middle to vein 2 near termen; a few bluish-white scales beyond the spots; a series of slight postmedian bluish-white streaks on veins 4 to 7; cilia white apically. Hind wings with some yellow scales on basal part of the media and vein 1; a yellow postmedial band from just above vein 6 to the anal margin, towards which it narrows; a yellow patch at apex. Length of forewing 21 – 23 mm.

Distribution: Australia (Queensland, New South Wales).

## Genus **Radinocera** HAMPSON

(fig. 33)

1910 *Radinocera* HAMPSON, Cat. Lepidopt. Phal. Brit. Mus., 9:457. – Species typica: *Phalaenoides maculosa* ROTHSCHILD (by monotypy).

Diagnosis: Proboscis fully developed; palpi porrect, reaching to just beyond frons; 2nd joint fringed with hair below; 3rd joint moderately long; frons with a large truncate conical prominence, with a raised rim at the extremity, and a corneous plate below; antennae ciliate; thorax mainly scaled, without crests, tibiae with moderate hair-fringe; abdomen smoothly scaled, with a basal scaly fan-shaped crest. Venation: in fore wings, vein 2 from about three-quarters of cell; 3, 4, 5 approximated, equidistant; discocellulars incurved; 6 from upper cell-angle; areole small; 7 and the short stalk of 8 + 9 from a point, at the extremity of areole; 10 from upper edge of areole, beyond the middle. Hind wings with vein 2 from five-sevenths of areole; 3, 4 and 6, 7 from the cell-angles; discocellulars incurved; 5 obsolescent from middle of discocellulars.

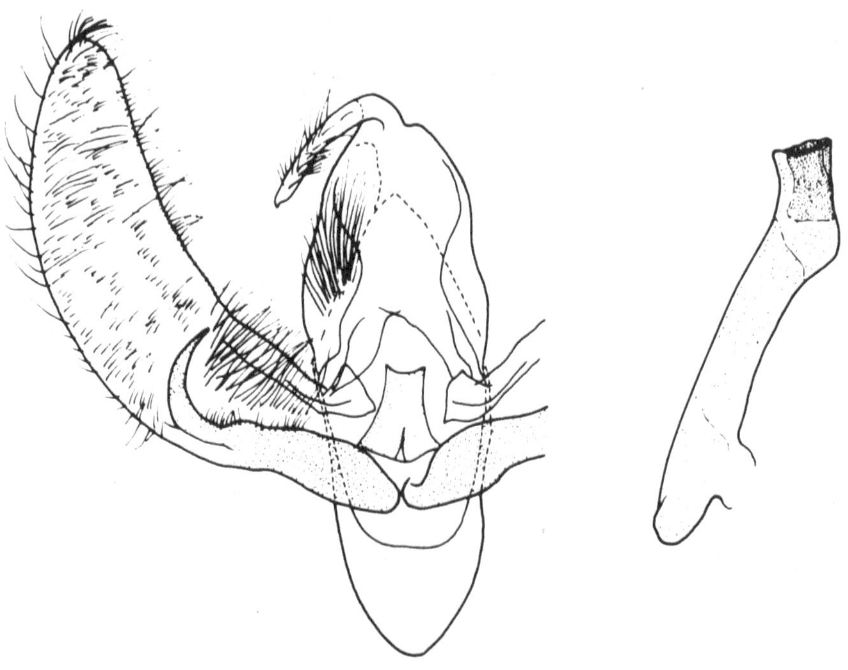

Fig. 33. *Radinocera maculosa* (ROTHSCHILD). Type. B. M. Noct. 8372

Male genitalia: Uncus relatively small, narrow, arched at the base only, then very faintly bent, ending in a minute hook. Vincula narrow, expanded at middle, broadening and fused distally; the expanded portion bearing a bunch of stiff hair. Valva elongate, banana-shaped; fold of sacculus rather narrow, reaching only to about one-quarter of sacculus, ending in a sickle-shaped harpe. Aedeagus about seven-tenths of costa in length, moderately robust, nearly straight, but elbowed at the distal quarter; vesica bearing minute cornuti; fultura inferior with the proximal angles produced. Saccus semi-elliptical.

Distribution: Indonesia, Australia, Tasmania.

Key to the species

1 Hind wings with a yellowish patch in proximal half . . . . . . . . . . . . . *R. placodes*, p. 143
1′ Hind wings with the proximal half white . . . . . . . . . . . . . . . . . . . . . . . . . . . . . . 2
2 In fore wings, distal edge of postmedial band rounded below the discal fold . . . . . . . *R. maculosa*, p. 143
2′ In fore wings, distal edge of postmedial band dentate below the discal fold . . . . . . . . *R. vagata*, p. 144

## Radinocera placodes (LOWER)

1903 *Cruria placodes* LOWER, Trans. R. Soc. South Australia, Adelaide, **27**:30 (♀). – Terra typica: Tasmania (Type, ♀, in South Australian Museum, Adelaide).
1910 *Radinocera placodes* – HAMPSON, Cat. Lepidopt. Phal. Brit. Mus., **9**:457; pl. 0, fig. 25.

Diagnosis: Body and wings black; head and thorax variegated with yellow; coxae yellow; abdomen with yellow edges to the urites. Fore wings with a yellow spot on dorsum and another on the discocellulars; an outer line consisting of two connected yellowish-white spots; apex with a square spot; tornus with a triangular spot. Hind wings with the basal part yellowish-white; a small apical spot and a small spot near the anal angle; cilia spotted with white and blackish. Length of fore wing 20 mm.

Remark: JORDAN in SEITZ (Großschmetterl. Erde, **11**:25, 1912) believes *R. placodes* to be a local form of *R. maculosa*. However, the disjunction between the distribution areas of both forms is considerable, and moreover *R. placodes*, being an insular form, has more probably reached the species-status.

Distribution: Tasmania.

## Radinocera maculosa (ROTHSCHILD)

1896 *Phalaenoides maculosa* ROTHSCHILD, Novit. Zool., **3**:94; pl. 15, fig. 29 (♂). – Terra typica: Cedar Bay, Queensland (Type, ♂, in British Museum).
1910 *Radinocera maculosa* – HAMPSON, Cat. Lepidopt. Phal. Brit. Mus., **9**:458; pl. 0, fig. 220.

Diagnosis: Head and body dark brown; palpi with the first and second joints banded with orange; head and thorax marked with orange; coxae and inner side of fore tibias orange; abdomen with broad orange edges to the urites. Fore wings with a large white spot occupying the second quarter of the dorsal area, and a small one beyond the middle; postdiscal area with a large double spot below costa, connected with a similar spot below; tornus with an elongate white spot; terminal area with partly rudimentary series of white points; cilia concolorous, tipped with yellowish apically and near tornus. Hind wings with a large basal white patch, and a small spot near the anal angle; cilia spotted with white and dark brown. Length of fore wing 18 mm.

Early stages: Larva whitish, with brown transverse bands; each segment with a ring of black warts bearing each a long whitish hair; head red; legs and a lateral spot on prolegs black.

Distribution: Queensland.

## Radinocera vagata (WALKER)

1864 *Nyctemera vagata* WALKER, List. Lepidopt. Het. Brit. Mus., **31**: 208 (♂). – Terra typica: North Australia (Type, ♂, in British Museum).
1900 *Phalaenoides kuhni* ROTHSCHILD, Novit. Zool., **7**:275; pl. 5, fig. 4 (♂). – Terra typica: Dammer Island (Type, ♂, in British Museum).
1910 *Radinocera vagata* – HAMPSON, Cat. Lepidopt. Phal. Brit. Mus., **9**:458.

Diagnosis: Head and thorax dark brown; 2nd joint of palpi white at base and at tip; head and thorax variegated with white; pectus with a white pilosity and orange spots; fore femora and tibiae with orange markings, rest of legs marked with white; abdomen orange, with the two proximal urites banded with white and black; underside orange, unbanded. Fore wings dark brown; white dots in basal area and in cell; dorsal area with a white patch in middle and a smaller one beyond; an irregular postcellular band, oblique, narrowing towards tornus; a series of subterminal dots; apical and tornal areas each with a white spot; cilia spotted black and white. Length of fore wing 18 – 22 mm.

Distribution: North Australia (terra typica); Dammer Island (terra typica of *Phalaenoides kuhni*); Tenimber, Letti, Soemba, Northeast Papua.

## Genus Coenotoca TURNER

(fig. 34)

1903 *Coenotoca* TURNER, Trans. R. Soc. South Australia, Adelaide, **27**:2. – Species typica: *Coenotoca monophyes* TURNER = *Melanochroia subaspersa* WALKER (by original designation).

Diagnosis: Proboscis fully developed. Antennae of male ciliate. Palpi obliquely upturned to just beyond frons; 2nd joint moderately scaled, 3rd moderate; frons with a pointed conical prominence; thorax clothed with hair, without crests; tibiae moderately fringed with hair; abdomen with dorsal crests on basal urites. Nervation: In fore wings, vein 2 given off at three-quarters of the cell; 3, 4, 5 approximated and equidistant; 6 from the lower cell-angle; areole short and

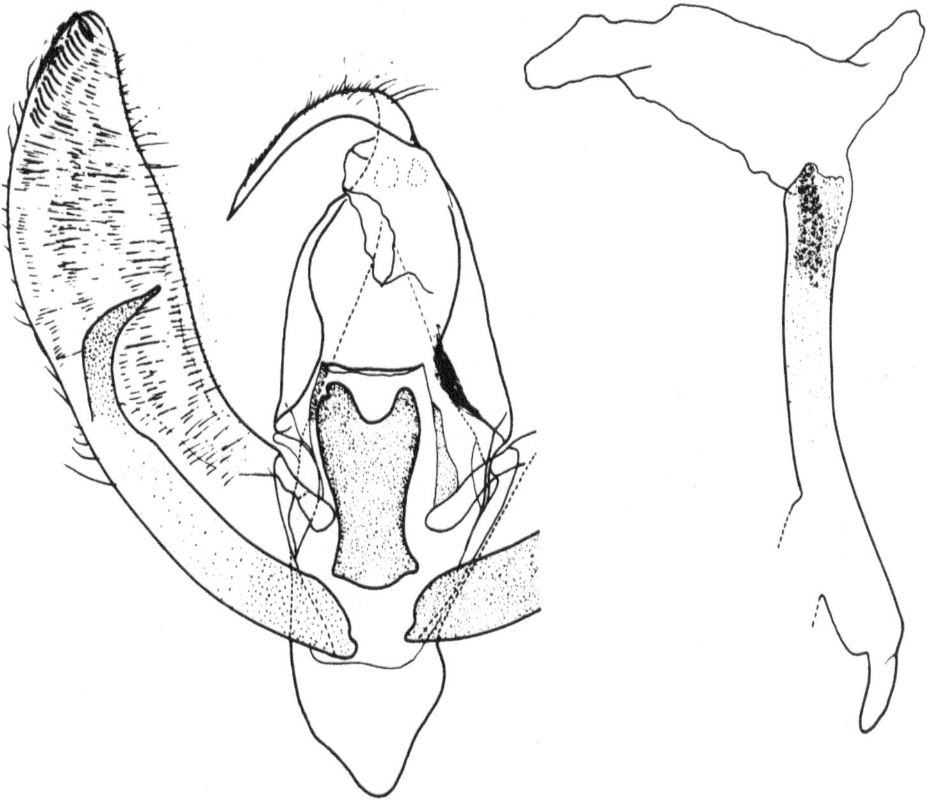

Fig. 34. *Coenotoca subaspersa* (WALKER). Type. B. M. Noct. 8380

narrow; 7 and 8 + 9 very short stalked, from the extremity of the areole; 10 from the upper side of the areole. In hind wings, vein 2 given off at three-quarters of the cell; 3, 4 and 6, 7 from the cell-angles; discocellulars incurved; 5 obsolescent, from middle of the discocellulars.

Male genitalia: Uncus slightly shorter than half of costa, narrow, compressed, bent at base, then straight, tapering to a point. Vinculum broadened in the median part; pilosity rather short and not forming true bunches. Valva banana-shaped; costa and sacculus more or less parallel; apex slightly rounded, forming an angle of about 50°; fold of sacculus not quite reaching half of valva, moderately broad, ending in a rather robust, regularly arched harpe, about two-thirds of uncus in length; a subapical series of stiff setae. Aedeagus nearly as long as the costa, rather slender, faintly arched; bearing a terminal longitudinal patch of minute cornuti; fultura inferior produced, with the lateral margins incurved proximally and excurved distally; distal margin produced into two lobes. Saccus slightly produced, rounded, with a somewhat narrowing terminal lobe.

Distribution: Australia.

## Key to the species

1 Abdomen with orange segmental lines; fore wings without blue markings; hind wings without blue markings; hind wings with a white band . . . . . . . . . . . *C. subaspersa*, p. 146
1' Abdomen with white segmental lines; fore wings with blue markings; hind wings with a series of yellowish-white spots . . . . . . . . . . . . . . . . . . . . . . . . *C. leonora*, p. 146

## Coenotoca subaspersa (WALKER)

1864 *Melanochroia subaspersa* WALKER, List Lepidopt. Het. Brit. Mus., **31**:210 (♂). – Terra typica: Moreton Bay, Queensland (Type, ♂, in British Museum).
1892 *Agarista fluitans* LUCAS, Proc. Linn. Soc. New South Wales, Sydney, (2) **8**:140 (♀). – Terra typica: Australia (Type, ♀, in South Australian Museum, Adelaide).
1903 *Coenotoca monophyes* TURNER, Trans. R. Soc. South Australia, Adelaide, **27**:2. – Terra typica: New South Wales (Type, ♂, in Australian Museum, Sydney).
1910 *Coenotoca subaspersa* – HAMPSON, Cat. Lepidopt. Phal. Brit. Mus., **9**:462–463; fig. 222.
1912 *Idalima subaspersa* – JORDAN in SEITZ, Großschmetterl. Erde, **11**:25; pl. 4c.
1912 *Idalima fluitans* – JORDAN in SEITZ, Großschmetterl. Erde, **11**:25.

Diagnosis: Body and fore wings very dark chocolate brown, nearly black; palpi with the second joint fringed with white below; frons barred with white; collar faintly edged with whitish; pectus with white pilosity; tibiae and tarsi ringed with white; abdominal crests on urites II and III orange at base; abdomen with slight orange yellow segmental lines, broader at sides; underside banded with white; terminal tuft black in female. Fore wings with an oblique white band from costa to beyond middle to termen just above tornus. Hind wings uniformly blackish brown. Length of fore wing about 15 mm.

Distribution: Queensland (Moreton Bay = terra typica), New South Wales.

## Coenotoca leonora (DOUBLEDAY)

1846 *Agarista leonora* DOUBLEDAY, Lord Stoke's Discovery of Australia, **1**:518; pl. 3, fig. 4. – Terra typica: Australia. – Terra typica restricta (JORDAN 1912): Port Essington, Northwest Australia (Type, ♂, in British Museum).
1892 *Phalaenoides leonora* – KIRBY, Syn. Cat. Lepidopt. Het.:23.
1910 *Idalima leonora* – HAMPSON, Cat. Lepidopt. Phal. Brit. Mus., **9**:459 – 460 (♂, ♀).
1912 *Argyrolepidia leonora* – JORDAN in SEITZ, Großschmetterl. Erde, **11**:22.

Diagnosis: Body and wings purplish black; palpi banded with pale yellow; sides of frons, basal antennal joint, spots on vertex, throat, edges of collar and tegulae yellow; coxae and femora with orange pilosity; tibiae and tarsi dotted with white; abdomen with segmental edges white, often on underside only; terminal urite orange. Fore wings with a whitish bar across cell at its end; an oblique series of three white spots beyond cell, two of them more distally above veins 3 and 4, the third more elongate, above vein 2; blue markings as follows: a basal spot; an oblique streak from base of dorsum to the oblique antemedial line; a short streak above cell-end; a spot below it; a postmedial series of small blue spots from costa to vein 2, then retracted to below cell-end and running as a line to dorsum; a curved subterminal series of blue spots; cilia white at apex and towards tornus. Hind wings with an oblique series of yellowish white spots from vein 4 beyond cell to tornus; cilia white apically and tornally and spotted with white in between. Length of fore wing 21 – 25 mm.

Distribution: North Australia, Queensland.

### Coenotoca leonora leonora (DOUBLEDAY)

1912 *Argyrolepidia leonora leonora* – JORDAN in SEITZ, Großschmetterl. Erde, **11**:22.

Diagnosis: As above.

Distribution: Northwest Australia (Port Darwin, Port Essington).

### Coenotoca leonora tasso (JORDAN)

1912 *Argyrolepidia leonora tasso* JORDAN in SEITZ, Großschmetterl. Erde, **11**:22; pl. 3g (♂). – Terra typica: Cedar Bay, Australia (Type, ♂, in British Museum).
1912 *Argyrolepidia leonora* aberr. *cedarensis* STRAND, Lepidopt. Cat., **5**:57. – Terra typica: Cedar Bay, Australia (Type, ♂, in British Museum).

Diagnosis: Hind wings with the whitish markings slightly broader, the most advanced spot is as large as it is far from termen.

Distribution: Queensland (Cedar Bay = terra typica).

## Genus **Periopta** TURNER

(fig. 35)

1920 *Periopta* TURNER, Trans. R. Soc. South Australia, Adelaide, **44**:127. – Species typica: *Agarista ardescens* BUTLER (by original designation).

Diagnosis: Antennae half to two-thirds of costa, filiform; frons with truncate conical prominence, with raised rim at apex; palpi moderate, ascending; the second joint with rather short, loosely appressed hairs; the third long, smooth; abdomen smooth, with dorsal crests more or less developed; hind tibiae with long hairs dorsally. Venation: in fore wings, vein 2 from about two-thirds of cell; 3, 4 from the lower cell-angle; 5 from well above the angle; 6 from the upper cell-angle; areole narrow; 7 and stalk of 8 + 9 from extremity of areole; 10 from the upper angle of areole. Hind wings with vein 2 from about three-quarters of cell; 3, 4 and 6, 7 from the cell-angles; 5 obsolescent, from middle of discocellulars. –

Male genitalia: Uncus short, slightly shorter than one-third of costa, compressed, arched at base, then nearly straight and bent at the extremity to form a minute hook. Tubus analis short and broad. Vinculum with the fold narrow, bearing a submarginal fringe of moderately long stiff hair. Valva elongate, rather narrow, banana-shaped; costa faintly outcurved at middle; apex upturned, more or less pointed;

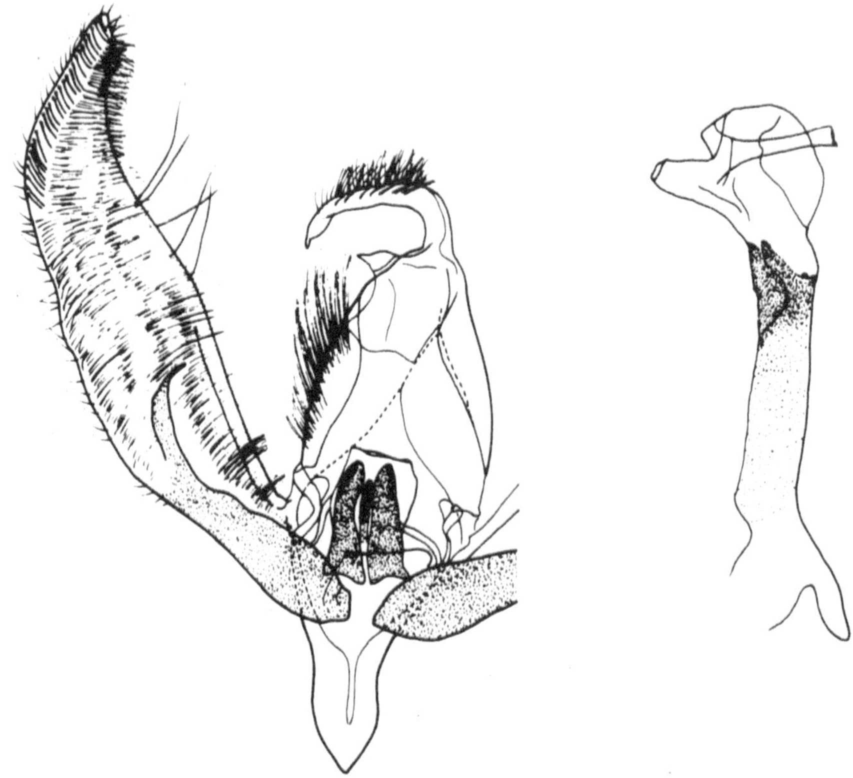

Fig. 35. *Periopta ardescens* (BUTLER). Type. B. M. Noct. 8374

sacculus outcurved at middle; apex upturned, more or less pointed; sacculus outcurved, but faintly incurved towards the end of the fold; latter moderate, reaching to slightly less than one-third of valva, ending in a short, arched harpe; terminal quarter of sacculus and apex with a submarginal row of short stiff spine-like hair. Aedeagus about two-thirds of valva in length, moderately robust, nearly straight, covered terminally with minute cornuti; fultura inferior consisting of a pair elongate triangular plates with distal narrow ends rounded and bearing a number of minute warts on the outer side; in between, an elongate narrow plate covered with minute warts. Saccus rather elongate, lance-shaped.

D i s t r i b u t i o n: Australia.

## Key to the species

1 Head black; spotted with yellow; abdomen black with a basal patch and the two last urites orange . . . . . . . . . . . . . . . . . . . . . . . . . . . . . . . . . . . . *P. ardescens*, p. 149

1' Head yellow; abdomen black banded with orange . . . . . . . . . . . . . . . *P. diversa*, p. 149

## Periopta ardescens (BUTLER)

1884 *Agarista ardescens* BUTLER, Ann. Mag. Nat. Hist., London, (5) **14**:405 (♀). – Terra typica: Port Darwin, North Australia (Type, ♀, in British Museum).
1892 *Phalaenoides ardescens* – KIRBY, Syn. Cat. Lepidopt. Het.:22.
1912 *Idalima ardescens* – JORDAN in SEITZ, Großschmetterl. Erde, **11**:25; pl. 3b.
1920 *Periopta ardescens* – TURNER, Trans. R. Soc. South Australia, Adelaide, **44**:127.

Diagnosis: Body and wings black; palpi with the first, base and extremity of the second, and base of the third joints pale yellow; sides of frons, occiput, collar, tegulae and thorax above striped with yellow; pectus, coxae, mid and hind femora and tibiae except at the extremity orange; fore legs and mid and hind tibiae white at the extremity; tarsi ringed with white; abdomen with a basal patch and the two last urites orange. Fore wings with the submedian area beyond the antemedial band and the terminal area suffused with purple; a yellowish-white basal maculate band; a subbasal point below costa; an oblique antemedial maculate band from below costa to vein 1; a square spot in cell-end; a band beyond cell from below costa to vein 2, oblique, excurved beyond each cell-angle and angled inwards in median fold; obliquely placed dots below it in the submedian fold and on vein 1; a postmedial series of points on veins from costa to vein 3; a subterminal series with larger spots on tornus; some silvery blue markings, viz.: in middle of cell; on the discocellulars; on base of vein 2; and above middle of vein 1; cilia dotted with white in male. Hind wings unspotted; cilia chequered with white and black. Length of fore wing 20 mm.

Distribution: North Australia (Port Darwin = terra typica), Queensland.

## Periopta diversa (WALKER)

1864 *Agarista diversa* WALKER, List Lepidopt. Het. Brit. Mus., **31**:49 (♀). – Terra typica: North Australia (Type, ♀, in British Museum).
1892 *Aegocera diversa* – KIRBY, Syn. Cat. Lepidopt. Het.:37.
1910 *Radinocera diversa* – HAMPSON, Cat. Lepidopt. Phal. Brit. Mus., **9**:457; pl. 147, fig. 24.
1912 *Idalima diversa* – JORDAN in SEITZ, Großschmetterl. Erde, **11**:25; pl. 4b.
1920 *Periopta diversa* – TURNER, Trans. R. Soc. South Australia, Adelaide, **44**:127.

Diagnosis: Body and wings blackish brown; head yellow; extremities of palpi, frontal prominence, patch on vertex brown; collar and tegulae orange, latter with black spots on shoulders; pectus pale yellow; mid and hind tibiae spotted with orange; abdomen with orange bands, narrow towards the base. Fore wings with some orange at base; an oblique orange antemedial bar from the subcosta to vein 1; a square spot in cell-end; an elliptical spot below base of vein 2; an oblique band beyond cell between veins 2 and 7, rounded at the extremities and constricted at middle; veins on terminal area with whitish streaks. Hind wings with a broad curved orange postmedial band from vein 7 to the anal margin, rounded above; cilia spotted brown and orange. Length of fore wing 18 mm.

Distribution: North and West Australia.

## Genus **Idalima** TURNER

(fig. 36)

1903 *Idalima* TURNER, Trans. R. Soc. South Australia, Adelaide, **27**:2. – Species typica: *Agarista affinis* BOISDUVAL (by original designation).

Diagnosis: Proboscis fully developed; palpi upturned, with the second joint reaching almost to middle of frons and fringed with hair in front; 3rd joint moderate, porrect; antennae of male ciliate; frons with a weak rounded prominence with a raised rim at the extremity and a corneous plate below it; thorax clothed with scales and hair mixed, without haircrests; abdomen smoothly scaled, without crests. Venation: in fore wings, vein 2 from two-thirds of cell; 3, 4, 5 separated, equidistant; upper discocellular faintly incurved; 6 from angle of cell; areole narrow; 7 and the short stalk of 8 + 9 from extremity of areole; 10 from upper angle of areole. Hind wings with vein 2 from three-fifths of cell; 3, 4 and 6, 7 from the cell-angles; discocellulars incurved; 5 obsolescent, from middle of discocellulars.

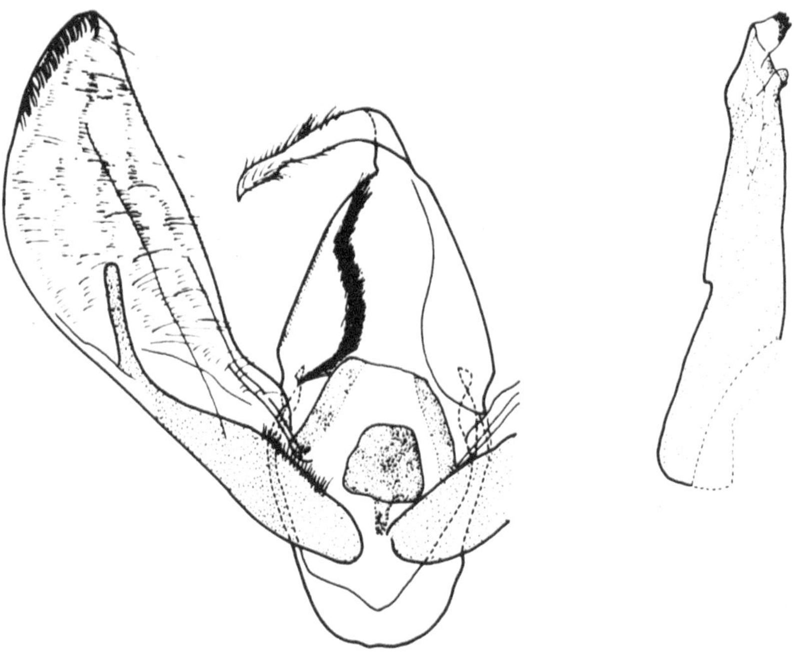

Fig. 36. *Idalima affinis* (BOISDUVAL). Type, B. M. Noct. 8379

Male genitalia: Uncus rather short, about two-thirds of vinculum in length, very faintly bent; terminal hook minute. Vinculum narrowing distally, with the inner margin regularly fringed with rather short hair. Valva moderately broad, rather narrow basally, with the broadest part at about two-thirds of its length; costa nearly straight; apex rounded, forming an angle of about 60°; termen and the distal part of sacculus forming a continuous arch; folded part of sacculus more or less straight; fold moderately broad, reaching to middle of valva and ending in a short, slender, almost straight harpe; a subapical and terminal row of short stiff hair. Aedeagus slightly longer than three-quarters of valva, robust, but slightly less so terminally, very faintly arched; a short subterminal and a slightly longer terminal processes, the latter bearing a number of minute cornuti; fultura inferior roughly semi-elliptical, with the distal margin rounded. Saccus broadly rounded.

D i s t r i b u t i o n: Australia.

Key to the species

1 Hind wings unmarked . . . . . . . . . . . . . . . . . . . . . . . . . . . . . . . . *I. affinis*, p. 151
1′ Hind wings with orange, yellow or white markings . . . . . . . . . . . . . . . . . . . . . . 2
2 Hind wings with a postmedial white band . . . . . . . . . . . . . . *I. hemiphragma*, p. 152
2′ Hind wings with a couple of small yellowish spots . . . . . . . . . . . . *I. metasticta*, p. 151

## Idalima affinis (BOISDUVAL)

1832 *Agarista affinis* BOISDUVAL, Voyage Astrolabe, Lépidopt.:177. – T e r r a t y p i c a: Australia (Type, ♂, in British Museum).
1884 *Agarista vindex* BUTLER, Ann. Mag. Nat. Hist., London, (5) 14:405 (♂). – T e r r a t y p i c a: Queensland (Type, ♂, in British Museum).
1892 *Phalaenoides affinis* – KIRBY, Syn. Cat. Lepidopt. Het.:22.
1892 *Phalaenoides vindex* – KIRBY, Syn. Cat. Lepidopt. Het.:23.
1903 *Idalima affinis* – TURNER, Trans. R. Soc. South Australia, Adelaide, 27:2.

D i a g n o s i s: Body black; first antennal joint, first and second joints of palpi at the extremity, sides of frons, throat, spots on occiput, collar, tegulae and on thorax, pale yellow; mid and hind coxae and outer side of mid tibiae with orange pilosity; fore coxae and tibiae with yellow pilosity; tarsi ringed with white; abdomen with white segmental lines; underside with pale yellow bands; anal tuft orange. Wings black, tinged with brown. Fore wings with two pale yellow basal dots; a subbasal series of three points, followed by a series of four spots; an antemedial series of four points, followed by a point in cell and a point on each side of vein 1; an oblique postmedian band from costa to the submedian fold; costa yellow from base to the band; a subterminal series of white points on veins; point on vein 5 more proximal; cilia yellow apically and at tornus. Hind wings unspotted; cilia spotted with pale yellow and black from apex to vein 2. Length of fore wing 16 mm.

D i s t r i b u t i o n: North Australia, Queensland, New South Wales.

## Idalima metasticta HAMPSON

1910 *Idalima metasticta* HAMPSON, Cat. Lepidopt. Phal. Brit. Mus., 9:461; pl. 146, fig. 27 (♂). – T e r r a t y p i c a: Adelaide River, North Australia (Type, ♂, in British Museum).

D i a g n o s i s: Body and wings black; palpi with the first and the extremity of the second joints yellowish white; sides of frons yellowish white; occiput, collar, tegulae and thorax spotted with pale yellow; throat and pectus yellowish white;

legs banded with yellowish white; mid tibiae with an orange patch above; abdomen banded with yellowish white; anal tuft orange at the extremity in female. Fore wings with two yellowish white basal spots; subbasal spots below costa and cell, fellowed by spots below costa, in and below the cell with an elongate mark on vein 1; antemedial spots below costa and on the media; sometimes a discocellular dot; conjoined spots beyond cell below costa and beyond the upper cell-angle, with two points beyond them; a larger spot beyond the lower cell-angle; a spot on vein 1 below cell-end; a more or less developed streak on base of dorsum; sometimes a slight streak on middle of dorsum; a subterminal series of small bluish-white spots on veins; spot on vein 5 more proximal; cilia white apically and tornally. Hind wings with yellowish streaks below cell and on dorsal area; spots beyond the discocellulars and the lower cell-angle, sometimes confluent and forming a band; cilia chequered white and black. Length of fore wing 11 mm.

Distribution: North Australia (Adelaide River = terra typica).

## **Idalima hemiphragma** (LOWER)

1903 *Cruria hemiphragma* LOWER, Trans. R. Soc. South Australia, Adelaide, **27**:30 (♂). – Terra typica: Perth, West Australia (Type, ♂, in South Australian Museum).
1910 *Idalima hemiphragma* – HAMPSON, Cat. Lepidopt. Phal. Brit. Mus., **9**:460; pl. 146, fig. 26.

Diagnosis: Body and wings dark red-brown; head and thorax mixed with some yellowish hair; a white postocular streak; coxae yellow; tarsi ringed with white; abdomen with the two last urites yellow, and the underside banded with yellow. Fore wings with a series of four white dots below the basal half of costa and above the basal half of vein 1; a small spot in middle of cell; a square spot in end of cell; a white postmedial band from costa to vein 2, its proximal edge angled outwards below costa and beyond the lower cell-angle, and its dorsal end incurved; a subterminal series of indistinct white dots; cilia white apically. Hind wing with a white band beyond cell from vein 6 to submedian fold; cilia white towards tornus. Length of fore wing 18 mm.

Distribution: West Australia (Perth = terra typica).

## Genus **Sarbissa**, genus **novum**

(fig. 37)

Diagnosis: Proboscis well developed; palpi oblique, the first and second joints with long, dense pilosity; the third slightly shorter than the second, naked, slender, compressed, but dilated terminally; frons with a truncate conical prominence; antennae reaching the last quarter of costa, filiform; pilosity of pectus and legs long and dense; outer spurs of the hind tibiae much shorter than the inner pair. Costa of the fore wing nearly straight; apex rounded; termen regularly rounded; tornus blunt. Nervation: in the fore wings, vein 2 very near the lower cell-angle; 3 slightly before the angle; 4 and 5 from the angle; areole small, rhomboidal; 6 from the upper angle of the cell; 7 and the stalk of 8 + 9 from the extremity of the areole; 10 from the upper margin of the areole. In the hind wings, vein 2 from very near the lower angle of the cell; 3 and 4 very much approximated; discocellulars very weak; 5 from middle of discocellulars, obsolescent; 6 and 7 from the upper angle of the cell; 8 anastomosing with the cell at one point shortly after base.

Differs from the genus *Sarbanissa* WALKER in the last palpal joint being dilated terminally, not uniformly slender; in vein 2 being placed in both wings very near the lower cell-angle; in vein 5 issuing in fore wing from the lower cell-angle together with vein 4.

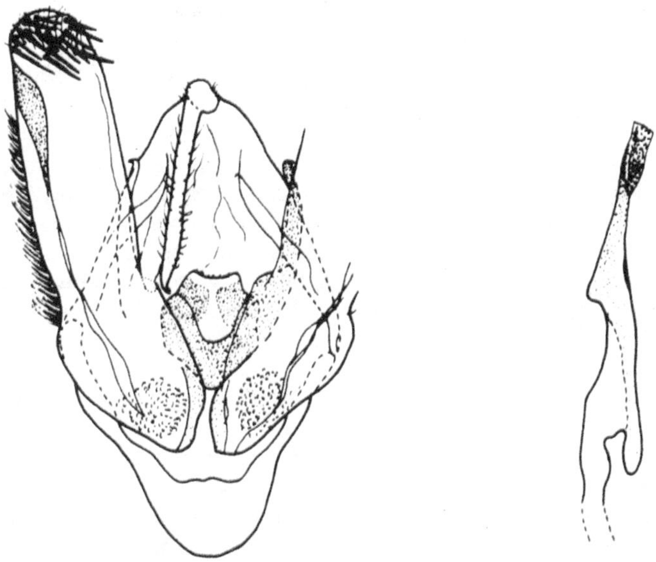

Fig. 37. *Sarbissa bostrychonota* (TAMS). Neallotype. B. M. Noct.

Male genitalia: Uncus as long as five-twelfths of valva, narrow, nearly straight, with sparse pilosity and a small terminal hook. Fold of vinculum narrow, without pilosity. Valva elongate, moderately broad; costa and sacculus nearly straight and parallel; apex broadly rounded, bearing a field of strong spines; harpe placed at one third of valva, slightly shorter than the uncus, slender, faintly arched, with a terminal hook. Aedeagus slightly longer than costa, slender; basal half spoon-shaped; terminal part and vesica covered with minute cornuti; fultura inferior broadly V-shaped, with the arms rounded. Saccus semi-elliptical.

Differs from *Sarbanissa* in lacking the usual bunch of stiff hair on the vinculum, and in the presence of a terminal area of strong, long spines on the valva; the shape of the valva also is different. The male genitalia show a surprising similitude with those of the madagascan genus *Ancarista* JORDAN, differing chiefly in the shape of the aedeagus. – Type species of the genus: *Seudyra bostrychonota* TAMS.

D i s t r i b u t i o n : Fiji Islands (Lautoka).

## Sarbissa bostrychonota (TAMS)

1929 *Seudyra bostrychonota* TAMS, The Entomologist, **62**:252–253 (♀). – Terra typica: Lautoka, Fiji Islands (Type, ♀, in British Museum).

Diagnosis: The male has not yet been described. The following description is from the neallotype, Fiji, Coloi Suva, August 1968 (H. S. & E. S. ROBERTSON), in the British Museum (Natural History).

Antennae black, sprinkled with white; palpi black, marked with white; head and collar dark chocolate brown, latter edged with whitish; thorax chocolate brown; pectus white, streaked with black in front; legs black, marked with white; hind femora and tibiae white, tinged above with orange; abdomen orange; basal crest grey brown; a dorsal dark brown stripe, widening distally; anal tuft pale grey brown; underside duller, marked with blackish brown. Fore wings with the dorsal area from the base to the outer line sepia brown with silky reflections, the dark colour continuing along the outer line and sending out proximally streaklets on veins to near the discocellular mark; a pure white spot below the typical markings, in interspaces Ia and II; costal and terminal areas white, strongly sprinkled with brown and grey; inner line undulate, triple, black, filled in with grey, its dorsal part below the anal vein merging into the dark colour of that area; typical markings white with a brownish tinge; outer line undulate, triple, filled in with sepia brown, rather strongly dentate on veins 2 and 3; subterminal line blackish, edged with white proximally; space before it browner; terminal line black; cilia dark grey spotted with white. Hind wings ochreous orange; a broad dark brown terminal band, wavy proximally; cilia blackish brown spotted with white. Length of the fore wing 22 mm.

Female: Rather different. Head yellowish white, collar yellow, both streaked with black; basal tuft of abdomen blackish brown, scales tipped with white. Fore wings lacking the dark brown dorsal area, and appearing much lighter coloured; typical marking larger and whiter. Length of fore wing 27 mm.

Distribution: Fiji Islands.

## Genus Sarbanissa WALKER

(fig. 38)

1865 *Sarbanissa* WALKER, List Lepidopt. Het. Brit. Mus., **33**:746. – Species typica: *Sarbanissa insocia* WALKER.

1875 *Seudyra* STRETCH, Cist. Entomol., **2**:19. – Species typica: *Eusemia transiens* WALKER.

1894 *Zalissa* HAMPSON (partim), Moths India, **2**:157. –

Diagnosis: Proboscis fully developed; antennae of male ciliated; palpi upturned, with the second joint reaching about to middle of frons and fringed with hair in front; third joint moderate or long; frons with a truncate conical prominence with a raised ridge at the extremity; thorax clothed with hair and hairlike scales and without crests; fore tibiae fringed with long hair; mid and hind tibiae moderately fringed with hair; abdomen with a dorsal series of hairy crests. Venation: in fore wings vein 2 from three-quarters of cell; 3, 4 much approximate; 5 well above the lower angle of cell; areole small; 6 from the upper cell-angle; 7 and stalk of 8 + 9 from tip of areole; 10 from the upper angle of areole. Hind wings with vein 2 from five-eights of cell; 3, 4 and 6, 7 from the cell-angles; discocellulars incurved; 5 obsolescent, from middle of discocellulars.

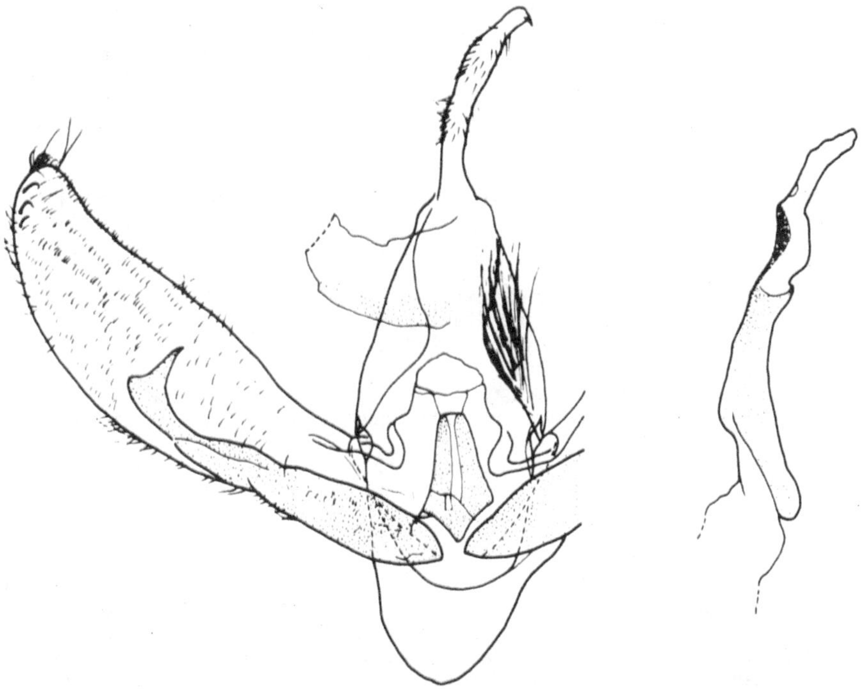

Fig. 38. *Sarbanissa insocia* WALKER. Type of *Phaegorista catocalina* WALKER. B. M. Noct. 8389

Male genitalia: Uncus moderately long, rather slender, somewhat bent, obtuse at the extremity, covered with short hair. Vincula with triangular expansion in the proximal half, bearing each a well-developed bunch of stiff hair. Valva elongate and narrow, somewhat banana-shaped; costa faintly incurved, but rather outcurved in the median portion; apex rounded; termen and sacculus evenly outcurved; fold of sacculus moderately broad, reaching to middle of valva and ending in a rather strong harpe bearing a slender transverse pointed process and a few stiff subapical setae. Aedeagus as long as the fold of valva plus the harpe, rather slender, arched in basal half, bearing subterminally a patch of minute cornuti, and ending in a slender, faintly arched process; fultura inferior triangular proximally, with the lateral margins converging distally, and the distal margin cut straight. Saccus broadly outcurved.

D i s t r i b u t i o n : East and South Asia, including the Philippine Islands and Indonesia.

## Key to the species

1 Palpi with the last joint long; hind wings with the termen slightly excised at submedian fold and the tornus lobed . . . . . . . . . . 2
1′ Palpi with the last joint moderate; hind wings with the termen evenly curved . . . . . . 3
2 Fore wings with postmedial line very oblique from costa to vein 4; hind wings without discocellular spot . . . . . . . . . . *S. venosa*, p. 156
2′ Fore wings with postmedial line nearly vertical, slightly incurved from costa to vein 4; hind wings with a black discocellular spot . . . . . . . . *S. subflava*, p. 157
3 Fore wings with the apex produced and the termen obliquely curved . . . . . . . . . . 4
3′ Fore wings with the apex not produced and the termen much less obliquely curved . . . 8
4 Hind wings without subterminal band . . . . . . . . *S. flavida*, p. 158
4′ Hind wings with a black subterminal band . . . . . . . . 5
5 Hind wings without discocellular spot . . . . . . . . *S. longipennis*, p. 159
5′ Hind wings with a black discocellular spot . . . . . . . . 6
6 Hind wings with the discocellular spot small . . . . . . . . 7
6′ Hind wings with the discocellular spot large . . . . . . . . *S. mandarina*, p. 159
7 Head dark reddish brown tinged with grey . . . . . . . . *S. insocia*, p. 160
7′ Head creamy white with sides of frons chocolate brown . . . . . . . . *S. vitalis*, p. 161
8 Both wings with the termen not crenulate . . . . . . . . 11
8′ Both wings with the termen slightly crenulate . . . . . . . . 9
9 Fore wings with a curved white patch beyond the cell . . . . . . . . 10
9′ Fore wings with a square white spot in end of cell . . . . . . . . *S. nepcha*, p. 161
10 Abdomen with the crests red brown . . . . . . . . *S. albifascia*, p. 161
10′ Abdomen with a single black crest . . . . . . . . *S. poecila*, p. 162
11 Hind wings with a square blackish spot above tornus, separate from the terminal band *S. venusta*, p. 162
11′ Hind wings without separate square black spot above tornus . . . . . . . . 12
12 Hind wings with the terminal band expanding into a large square spot above tornus . . . 13
12′ Hind wings not expanding into a large square spot above tornus . . . . . . . . 15
13 Fore wings with the costal area and cell suffused with white to postmedial line and thence obliquely to termen above tornus . . . . . . . . *S. interposita*, p. 163
13′ Fore wings without white suffusion . . . . . . . . 14
14 Abdomen black banded with yellow . . . . . . . . *S. bala*, p. 164
14′ Abdomen orange with a dorsal series of transverse black spots . . . . . . . . *S. cirrha*, p. 165
15 Abdomen with only two basal tufts and the extremity black; hind wings with the terminal band very narrow . . . . . . . . *S. jordani*, p. 165
15′ Abdomen with a series of black dorsal spots . . . . . . . . 16
16 Hind wings without a black discocellular spot . . . . . . . . *S. transiens*, p. 166
16′ Hind wings with a black discocellular spot . . . . . . . . 17
17 Hind wings with the postmedian band at least 3 mm wide, yellowish white . . . . . . . . *S. catocaloides*, p. 167
17′ Hind wings with the postmedian band only about 1 mm wide, pale creamy yellow *S. melanura*, p. 167

## Sarbanissa venosa (MOORE)

1879 *Seudyra venosa* MOORE, Proc. Zool. Soc. London, **1879**:389 (♂, ♀). – Terra typica: Northeastern India (Type, ♂, in British Museum).
1894 *Zalissa venosa* – HAMPSON, Moths India, **2**:156 (♂).

Diagnosis: Head and thorax dark chocolate red brown, mixed with some greyish ochreous; collar edged with ochreous; pectus and legs orange; fore and mid tibiae sprinkled with brown; abdomen orange with blackish brown crests. Fore wings deep chocolate red brown; costal half to postmedial line thickly sprinkled with ochreous, extending to just below cell and vein 2, and beyond the postmedial line to termen between vein 4 and tornus; the veins striated with ochreous except vein 1; dorsal area deep red with some purplish-violet on middle of dorsum and beyond the postmedial line; apical area deep red to vein 5 with a purple-violet patch below apex and another between veins 3 and 5 beyond the postmedial line; subbasal line present as a slight ochreous striga from cell; antemedial line obsolete on costal half, vertical and ochreous below cell, with some purple-violet margined dorsally by a white line in submedian interspace; cell- and discocellular spots large, edged with ochreous; the former elongate elliptical, just touching the latter, which is broad and oblique; postmedial line double, sinuous, ochreous edged with blackish proximally; subterminal line fine, whitish, interrupted; a terminal series of small dark lunules; cilia ochreous with a dark basal line; apical area mixed with dark. Hind wings orange; terminal area black, broad at costa, in submedian fold angled inwards and followed by two dark lines with an orange line between them to near tornus; above, a triangular deep red brown patch; cilia ochreous with a dark line near the base. Length of fore wing 22 – 27 mm.

D i s t r i b u t i o n: Northeastern India to the Shan States.

### Sarbanissa subflava (MOORE)

1877 *Seudyra subflava* MOORE, Ann. Mag. Nat. Hist., London, (4) **20**:85 (♂). – T e r r a t y p i c a: Kiukiang, Central China (Type, ♂, in British Museum).

D i a g n o s i s: Head and thorax dark chocolate brown mixed with ochreous; ends of tegulae and metathorax with metallic blue gloss; palpi with the second joint dark laterally; pectus and legs orange; fore and mid tibiae ochreous mixed with dark brown; abdomen orange with a dorsal series of dark spots; crests brown mixed with blue. Fore wings deep chocolate brown; costal half to postmedial line thickly sprinkled with ochreous, extending to the submedian fold and beyond the postmedial line extending to termen between vein 3 and tornus; veins in dorsal half and submedian fold streaked with ochreous; dorsal and postmedian areas chocolate-red with some violaceous on media, on dorsum and before the subterminal line from below costa to vein 3; subbasal line faint, oblique, ochreous, from costa; antemedial line ochreous, obsolescent, oblique from costa to middle of cell, then vertical partly edged with white proximally and with blackish distally; cell-spot smallish, elongate-elliptical; discocellular spot large, both with an ochreous ringlet; postmedial line ochreous, double, to below cell where it is single, edged with blackish proximally; subterminal line bluish-white, undulate, crossed by a wedge-shaped blackish mark edged with ochreous above vein 5 from near termen, below vein 3 ochreous, angles inwards at vein 2, finally bluish-white again; terminal area sprinkled with ochreous except towards apex; with fine undulate ochreous line before termen, with dark striae beyond it; cilia ochreous, mixed with brown and with a dark median line. Hind wings orange; terminal band broad, brown tinged with red except towards apex; above tornus a large orange lunule; a small black discocellular spot; an undulate orange line before termen from vein 4 to anal angle; cilia white with dark lines near base and at tips. Length of fore wing 19 – 21 mm.

D i s t r i b u t i o n: Japan, Amurland, China, Korea.

E a r l y s t a g e s : Larva white except on tergites 1 and 11; these, head and sides of all segments ochreous brown; head and pronotum with two transverse series of black dots; segmental incisions greyish-black; transversal bands forming rings of conical warts; on the white dorsal areas four warts; each wart bearing one or two black setae, tipped with whitish; legs black; prolegs black on the outer side. On *Vitis* (*pachyphylla*?).

### S a r b a n i s s a s u b f l a v a s u b f l a v a (MOORE)

D i a g n o s i s : As above.

D i s t r i b u t i o n : China (Manchuria, Shantung, Chekiang, Kansu, Kiangsi, Hupé, Kwangtung, Taiwan); also Korea.

### S a r b a n i s s a s u b f l a v a j a n k o w s k i i (ALPHERAKY)

1907 *Zalissa jankowskii* ALPHERAKY, Rom. Mém., **9**:151; pl. 11, fig. 2 (♂, ♀). – T e r r a t y p i c a : Amurland (Types, ♂, ♀, in Leningrad Museum?).
1939 *Seudyra subflava jankowskii* – BYTINSKI–SALZ, Entomol. Rec., **51**:120; pl. 9, fig. 14.

D i a g n o s i s : Variable, but on the underside of the hind wings the submedian band obliterated or completely wanting.

D i s t r i b u t i o n : Amurland.

### S a r b a n i s s a s u b f l a v a j a p o n i c a (BYTINSKI–SALZ)

1939 *Seudyra subflava japonica* BYTINSKI–SALZ, Entomol. Rec., **51**:120 (♂, ♀). – T e r r a t y p i c a : Hakone, Japan (Type in collection of BYTINSKY–SALZ).

D i a g n o s i s : Abdomen with the dorsal dark spots well developed on all segments not only on the first 3 or 4; fore wings much less ochreous; only the veins, the edges of the cell- and discocellular spots and postmedial line narrowly yellow; discal spot on the hind wings much larger, and terminal band much wider; anal orange patch lies inside the band and is completely separated from the yellow discal area.

D i s t r i b u t i o n : Japan (Terra typica = Hakone near Fuji).

### Sarbanissa flavida (LEECH)

1890 *Seudyra flavida* LEECH, Entomologist, **23**:110 (♂, ♀). – T e r r a t y p i c a : Changyang, Western China, (Types, ♂, ♀, in British Museum).

D i a g n o s i s : Head and thorax purple-black; pectus and legs orange; tibiae and tarsi with some fuscous scales; abdomen orange with slight dorsal series of fuscous points. Fore wings whitish tinged with ochreous, strongly sprinkled with black; dorsal area chocolate-red with a leaden fascia at base, leaden bands before the antemedial and beyond the postmedial fulvous lines and a large leaden patch between them, the distal edge rounded before tornus; cell-spot elliptical; discocellular mark very large; both blackish, edged with whitish, with leaden centres and ochreous inner circles; postmedial line edged with black proximally; distally, a large apical chocolate-red patch with the centre purplish-leaden, edged distally by an obscure dentate white line; a terminal series of dark striae. Hind wings orange with a slight black discocellular spot and traces of an irregular subterminal line. Length of fore wing 26 – 30 mm.

Distribution: China: South Hupé (Changyang = terra typica), Szechuan, South Shensi.

## Sarbanissa longipennis (WALKER)

1865 *Catocala longipennis* WALKER, List Lepidopt. Het. Brit. Mus., **33**:934 (♂). – Terra typica: Silhet (Type, ♂, in British Museum).
1892 *Seudyra longipennis* – KIRBY, Syn. Cat. Lepidopt. Het.:38.
1894 *Zalissa longipennis* – HAMPSON, Moths India, **2**:156.

Diagnosis: Head and thorax dark chocolate-red-brown, with some silvery blue gloss; collar with a yellow and red line at base; pectus and legs orange yellow; fore tibiae sprinkled with red and fringed with greyish hair; abdomen orange; dorsal crests black. Fore wings with the costal half to postmedial line, the media and vein 2, and thence obliquely to termen at vein 4 and tornus, yellow, sprinkled with chocolate-red; area below cell to postmedial line deep chocolate-red; dorsal area from above vein 1 leaden grey; apical area chocolate-red to vein 5, with a leaden grey patch from below costa to below vein 6; antemedial line obsolete on the costal half, then yellow, edged proximally with leaden grey, and distally with black; cell-spot small, elliptical, with a yellow centre edged with red and ochreous; discocellular mark very large, fabiform, with a yellowish centre, edged irregularly by chocolate-red, with a central reddish lunule, some silvery suffusion on veins, and an ochreous inner ring; postmedial line indistinct, double, filled in with yellowish and edged with yellowish distally, angled outwards on vein 4, edged proximally by black below the submedian fold, and with a leaden grey band on the distal side, and followed then by a deep Indian red triangular spot; a subterminal series of deep Indian red striae, accompanied by some yellow in the apical area; terminal line fine, brown; cilia brown and yellowish. Hind wings orange-yellow, with an irregular brown subterminal band, with the distal edge slightly dentate at the veins, not reaching tornus and the anal margin; a series of terminal orange-yellow lunules. Length of fore wings 29 – 32 mm.

Distribution: Sikkim, Bhutan, Assam, Silhet (terra typica).

## Sarbanissa mandarina (LEECH)

1890 *Seudyra mandarina* LEECH, Entomologist, **23**:110 (♂). – Terra typica: Changyang, Western China (Type, ♂, in British Museum).

Diagnosis: Head and thorax dark brown, mixed with greyish-ochreous; pectus and legs ochreous-yellow; tibiae and tarsi sprinkled with brown; abdomen orange, sprinkled with black dorsally, and with black crests. Fore wings with the costal half to the postmedial line, extending to just below cell and vein 2, dark brown, thickly sprinkled with ochreous; the media and veins given off from it pale; dorsal and terminal areas purplish-violet, with deep red patches before the antemedial and beyond the postmedial lines; apical area red to vein 4, with a violet patch at middle; antemedial line obsolete on costal area, double below cell; cell spot and discocellular mark large, with dark chocolate centres and slight ochreous inner rings, and with a violet-white suffusion; postmedial line double; traces of a diffused white subterminal line; a fine ochreous line just before termen; cilia reddish brown. Hind wings orange-yellow; discocellular spot large, round, black; the submedian interspace covered with black hair to the very irregular black subterminal band, expanding into patches in intervals I and V, and excurved at middle.

D i s t r i b u t i o n: China (South Hupé [Changyang = terra typica]).

## Sarbanissa insocia WALKER

1865 *Sarbanissa insocia* WALKER, List Lepidopt. Het. Brit. Mus., **33**:746. – T e r r a t y p i c a: Sikkim (Type, ♂, in British Museum).
1866 *Phaegorista catocalina* WALKER, List Lepidopt. Het. Brit. Mus., **35**:1859 (♂, ♀). – T e r r a t y p i c a: Sikkim (Type, ♀, in British Museum).
1892 *Seudyra catocalina* – KIRBY, Syn. Cat. Lepidopt. Het.:38.
1892 *Zalissa exiguifascia* SWINHOE, Cat. Het. Mus. Oxon., **1**:167. – T e r r a t y p i c a: Sikkim (Type, ♂, in Museum Oxford University).

D i a g n o s i s: Head and thorax dark reddish-brown with bluish-grey reflections; base of palpi orange; collar with an orange basal line and a blue-grey edge; fore legs grey mixed with brown; pectus, and mid and hind legs orange-yellow; mid tibiae with some brown; abdomen orange with the dorsal crests black. Fore wings with the costal half to postmedial line, extending to the media and vein 2, dark brown, thickly sprinkled with yellow; submedian interspace to subterminal line, and apical area to vein 4 deep carmine-red; dorsal and terminal area below vein 4 brownish, suffused with blue-grey; base of the media and vein 1 beyond the antemedial line with slight white streaks; antemedial line yellow, obliquely curved, edged at both sides with brownish, and with a whitish line before it from cell to vein 1; cell spot and discocellular mark large, blackish with a white circle; the former round, with the centre edged with yellow; the latter oblique, with the centre defined by bluish-white; postmedial line double, yellow, incurved from costa to vein 4, where it is strongly angled outwards, edged in dorsal area with brown proximally and with whitish distally; apical area with a blue-grey patch from below costa to below vein 6; subterminal line whitish, undulate from vein 4 to tornus; a slight undulate yellow line just before termen, which is deep chocolate-red; cilia blackish-brown with a fine yellowish basal line. Hind wings deep orange; a small black discocellular spot; subterminal band very irregular, between veins 1 and 6. Length of fore wing 25 – 27 mm.

D i s t r i b u t i o n: Sikkim, Bhutan.

## **Sarbanissa vitalis** (JORDAN)

1926 *Seudyra vitalis* JORDAN, Novit. Zool., **33**:377 – 378 (♂). – Terra typica: Thado Cuarao, Indo-China (Type, ♂, in British Museum).

Diagnosis: Head creamy white; palpi blackish-brown, with the first joint below, and the extremities of the second and the third white; sides of frons and a postocular streak chocolate-brown; thorax chocolate-brown with three white stripes; abdomen, pectus, and legs buffy yellow; abdomen buffy yellow with a dorsal row of rather diffuse black spots; fore and mid tibiae and tarsi blackish. Fore wings creamy white; costal area blackish-brown from base to two thirds, emitting two short short chocolate-brown bars, one oblique across cell to just below base of vein 2, the other including the discocellulars, 3 mm broad, reaching to vein 2 and emitting three streaks on veins 2, 3 and 4; a terminal band of chocolate-brown lines and spots separated by creamy white interspaces, proximally well defined but with the components more or less diffuse, especially towards dorsum where there is more of white than of dark; from base along the dorsum, a chocolate-brown stripe, slightly narrowing distally, joining the terminal band, and bearing a diffuse white line. Hind wings buffy yellow; a rounded spot on the upper cell-angle, another slightly larger, somewhat distally from the lower cell-angle which it does not touch; terminal band from the costal margin to near vein 4, scalloped proximally, followed by a submarginal row of three spots; markings of hind wings chocolate-brown. Length of fore wing 31 mm.

Distribution: Vietnam (Thado Cuarao = terra typica).

## **Sarbanissa nepcha** (MOORE)

1867 *Catocala nepcha* MOORE, Proc. Zool. Soc. London, **1867**:68. – Terra typica: Moulmein, Burma (Type, lost).
1894 *Zalissa* (?) *nepcha* – HAMPSON, Moths India, **2**:157.
1910 *Seudyra nepcha* – HAMPSON, Cat. Lepidopt. Phal. Brit. Mus., **9**:440.

Diagnosis: Head black; antennae yellow; thorax rufous-brown; abdomen orange yellow with a dorsal and lateral series of blackish-brown spots. Fore wings rufous-brown; some irregular blackish marks on basal area; a square white spot in cell-end; a narrow slightly curved postmedial band edged with black proximally, and a series of dentate black marks beyond it; a terminal series of black and grey lunules; cilia grey brown. Hind wings orange-yellow with a broad black terminal band. Length of fore wing 32 mm.

Distribution: Sikkim.

## **Sarbanissa albifascia** (WALKER)

1865 *Catocala albifascia* WALKER, List Lepidopt. Het. Brit. Mus., **33**:933 (♀). – Terra typica: Moulmein, Burma (Type, ♀, in British Museum).
1894 *Zalissa albifascia* – HAMPSON, Moths India, **2**:157.
1910 *Seudyra albifascia* – HAMPSON, Cat. Lepidopt. Phal. Brit. Mus., **9**:439 – 440; fig. 209.

D i a g n o s i s : Head and thorax dark chocolate-brown mixed with grey and white; pectus and legs orange; fore tibiae and tarsi brown ringed with white; abdomen orange with red-brown crests. Fore wing deep chocolate-red; dorsal and terminal areas suffused with purplish-grey, and with a few white scales; a curved white patch beyond cell from below costa to vein 2, with some olive-ochreous suffusion round it; another patch of olive-ochreous suffusion from base to costa, above and below the media to the antemedial line, which is indistinct, red, double, undulate; a round red cell-spot; a discocellular red bar; postmedial line indistinct, double, red, filled in with olive-ochreous, and with white at costa, finally undulate; a slight whitish mark beyond it on costa; subterminal line indistinct, pale, diffuse, irregularly dentate, oblique in interval III and incurved at vein 2; a lunulate red terminal line; cilia dark brown with a whitish basal line. Hind wings orange; terminal area blackish-brown, broadest at costa; a white apical patch; cilia tipped with white to below vein 3, then with a white median line. Length of fore wing 21 – 28 mm.

D i s t r i b u t i o n : India, South China (Kwangtung coasts), Burma (Moulmein = terra typica), Malaya, the Andaman Islands.

## **Sarbanissa poecila** (JORDAN)

1912 *Seudyra poecila* JORDAN in SEITZ, Großschmetterl. Erde, **11**:27 (♂). – T e r r a t y - p i c a : Sikkim (Type, ♂, in British Museum).

D i a g n o s i s : Head and thorax dark chocolate-brown mixed with grey and white scales; prothorax purplish-red; abdomen orange with a single black crest. Fore wings bluish-white; basal and marginal areas and before dorsum purplish-red; beyond the cell, a large white patch margined with black; a postmedial bluish-grey line, becoming white towards costa. Hind wings yellow; a black terminal band, narrowing towards the anal margin, and not reaching tornus; cilia spotted with white. Length of fore wing 13 mm.

D i s t r i b u t i o n : Sikkim (?).

## **Sarbanissa venusta** (LEECH)

1888 *Sarbanissa venusta* LEECH, Proc. Zool. Soc. London, **1888**:614; pl. 31, fig. 2 (♂). – T e r r a t y p i c a : Gensan, Korea (Type, ♂, in British Museum).

D i a g n o s i s : Head and thorax dark brown mixed with white; pectus and legs mostly white; abdomen orange with black dorsal crests. Fore wings whitish sprinkled with blackish-brown; cell spot and discocellular mark large, black, sprinkled with white; dorsal area with a leaden-violaceous patch with some bright red at each extremity, extending from near base to near tornus, traversed by the pale ante- and postmedial lines; the former double, edged with blackish-brown distally, the latter double; a large apical black patch with a bright red spot on it; leaden spots with whitish marks on them in intervals III and V; an undulate whitish line close to termen. Hind wings orange with a black discocellular spot; a blackish-brown terminal patch from costa to vein 2, narrow and dentate towards vein 2, and a large spot above tornus. Length of fore wing 18 – 21 mm.

Distribution: Korea (Gensan = terra typica), China (Chekiang, Changyang, Szechuan, to Northwestern Yunnan).

### Sarbanissa venusta venusta (LEECH)

1936 *Seudyra venusta venusta* – MELL, Stettin. Entomol. Ztg., 97:182 – 183; fig. 18.

Diagnosis: Hind wings deep chrome-yellow to light orange.

Distribution: Species' area except Yunnan.

### Sarbanissa venusta yunnana (MELL)

1936 *Seudyra venusta yunnana* MELL, Stettin. Entomol. Ztg., 97:183. – Terra typica: Northwestern Yunnan, China (Type, ♂, in Berlin Museum).

Diagnosis: Hind wings ochreous-brown (capucine yellow); greyish-black terminal band broader, darker; dark cell-spot round.

Distribution: Northwest Yunnan.

### **Sarbanissa interposita** (HAMPSON)

1910 *Seudyra interposita* HAMPSON, Cat. Lepidopt. Phal. Brit. Mus., 9:437; pl. 148, fig. 11 (♀). – Terra typica: Chin-fu-san, Western China (Type, ♀, in British Museum).

Diagnosis: Head and thorax greyish-ochreous, mixed with some black; metathorax with some blue scales; palpi black with the extremities of joints white; pectus, legs and abdomen orange, the latter with dorsal series of black spots. Fore wings with the costal half to the postmedial line and thence obliquely to termen above tornus, white sprinkled with dark brown except before the postmedial line from below costa to vein 3; the dorsal and apical areas deep chocolate-red with patches of violet-grey on dorsal medial area and below apex; subbasal line oblique, double, black, from costa; a black point in cell before the antemedial line, which is obsolete to the media, then double, ochreous, filled in with violet-red, vertical; cell spot and discocellular mark with white inner rings, their centres sprinkled with black and silvery, the former round, the latter fabiform; postmedial line double, ochreous white, filled in with violet towards dorsum, oblique, incurved from costa to vein 4 near the subterminal line, incurved below vein 2; subterminal line indistinct, whitish, undulate, angled inwards in distal fold, obsolete towards costa; a terminal series of black striae slightly defined proximally by whitish; cilia blackish, with fine yellowish basal and medial lines. Hind wings orange; a small black discocellular spot; terminal black band broadly rounded on apical area, squarely dilated towards tornus; cilia whitish at the tip, yellow towards tornus. Length of fore wing 17 – 21 mm.

Distribution: China.

### Sarbanissa interposita interposita (HAMPSON)

1936 *Seudyra interposita interposita* – MELL, Stettin. Entomol. Ztg., **97**:181; fig. 16.

Diagnosis: Fore wings with the underside much darker than in *S. venusta*; white postmedial line broadly conjoined with the tornal spot; hind wings on the underside with the terminal band grey, washed out towards the anal margin.

Distribution: West and Central China.

### Sarbanissa interposita kuantungensis (MELL)

1936 *Seudyra interposita kuangtungensis* MELL, Stettin. Entomol. Ztg., **97**:182; fig. 17. – Terra typica: Kwangtung, China (Type, ♂, in Berlin Museum).

Diagnosis: Fore wings with the dark submarginal band distinctly narrowing towards tornus, almost reaching the latter; termen less arched; costa straighter; on the underside, the white postmedial band separeted from the white tornal spot by a narrow black streak that connects the black dorsal area with the black submarginal area; hind wings on the underside with the terminal band broader, blacker.

Distribution: Kwangtung.

### **Sarbanissa bala** (MOORE)

1865 *Phaegorista bala* MOORE, Proc. Zool. Soc. London, **1865**:795 (♂, ♀). – Terra typica: Sikkim (Types, ♂, ♀, in British Museum).
1894 *Zalissa bala* – HAMPSON, Moths India, **2**:156.
1895 *Seudyra bala* – KIRBY, Syn. Cat. Lepidopt. Het.:38.

Diagnosis: Head and thorax dark brown with some grey hair; pectus and legs orange; fore legs with brown markings; mid and hind tarsi sprinkled with brown; abdomen orange with dorsal series of black bands. Fore wings dark brown, sprinkled with grey-ochreous, except dorsally and apically; costa sprinkled with whitish; dorsal area with rusty-red spots near base and towards tornus; tornal spot round and edged with whitish; area below the media and apical area suffused with reddish; antemedial line only present below cell, indistinct, grey-ochreous; cell spot and discocellular mark large, dark brown with greyish-ochreous inner rings; the former round, the latter obliquely elliptical and extending to well below cell; postmedial line indistinct, greyish, oblique, incurved from costa to vein 4, then undulate; beyond, a faint oblique straight line from costa to vein 4; subterminal line faint, whitish; cilia dark brown, with the base pale greyish. Hind wings orange; a small black discocellular spot; terminal band blackish-brown, extending at costa to middle, strongly excurved at middle of termen, not quite reaching the anal margin; cilia tipped with white. Length of fore wing 20 – 22 mm.

Distribution: Mountains of Northern India, Burma.

### Sarbanissa bala bala (MOORE)

1912 *Seudyra bala bala* – JORDAN in SEITZ, Großschmetterl. Erde, **11**:27; pl. 4 d.

Diagnosis: As above.

Distribution: Sikkim (7000 – 10000 ft.), Bhutan.

### Sarbanissa bala hilaris (JORDAN)

1912 *Seudyra bala hilaris* JORDAN in SEITZ, Großschmetterl. Erde, **11**:27 (♀). – Terra typica: Manipur, Burma (Type, ♀, in British Museum).

Diagnosis: Abdomen with the black bands broader; markings of fore wings much more distinct; discal band whitish, at least on the disk, with distally a white line, and some white scales before termen.

Distribution: Burma (Manipur, Mao, 5000 – 7000 ft.).

### Sarbanissa cirrha (JORDAN)

1912 *Seudyra cirrha* JORDAN in SEITZ, Großschmetterl. Erde, **11**:27 (♂). – Terra typica: Taiwan, China (Type, ♂, in British Museum).

Diagnosis: Very near *Sarbanissa bala*. Abdomen orange, with a dorsal series of transverse black spots and the extremity black. Fore wings nearly as light coloured in cell and on disk as in *S. bala hilaris*; blackish-brown terminal band dentate on proximal margin between veins $R_1$ and $R_3$, very narrow between $R_3$ and $M_2$, nearly interrupted, and forming a large patch before tornus; a yellow space between the band and the cilia.

Distribution: China (Taiwan).

### Sarbanissa jordani (CLENCH)

1953 *Seudyra jordani* CLENCH, Ann. Carnegie Mus., Pittsburgh, **33**(6):143. – Terra typica: Mindanao, Philippine Islands (Type, ♂, in Carnegie Museum).

Diagnosis: Differs from *S. transiens dissimilis* SWINHOE as follows: slightly larger; fore wings more elongate; abdomen with only two basal urites tufted; terminal urite black. In fore wing, the red-brown colour somewhat paler; the large metallic blue spot on middle of dorsum rather large. In hind wings, the black border much narrower, less than half as wide as in *S. t. dissimilis*, with its proximal margin quite straight, not irregular.

Distribution: Philippine Islands (Mindanao).

## Sarbanissa transiens (WALKER)

1856 *Eusemia transiens* WALKER, List Lepidopt. Het. Brit. Mus., 7:1588 (♂, ♀). – Terra typica: Java (Types, ♂, ♀, in British Museum).
1875 *Seudyra transiens* STRETCH, Cist. entomol., London, 2:19.
1894 *Zalissa transiens* HAMPSON, Moths India, 2:155; fig. 102.

Diagnosis: Head and thorax dark brown mixed with grey; collar with a darker median line; fore legs grey, mixed with brown; pectus, mid and hind legs orange-yellow, mid tibiae sprinkled with brown; abdomen orange-yellow with black dorsal spots; terminal urite in female woolly, sometimes blackish. Fore wings dark brown, sprinkled with greyish-ochreous on basal area, on costal area to the postmedial line, on the veins and on terminal area above tornus; a more or less developed oblique diffused white band beyond cell from costa to vein 2, sometimes connected with a tornal patch; a subbasal deep Indian red spot above vein 1, with a smaller patch below it on tornus; antemedial line indistinct, ochreous, oblique from costa to below cell, then vertical, edged on both sides with Indian red from cell to dorsum, and with a leaden-grey band in costal area; medial area with an Indian red patch below cell and a leaden-grey patch on dorsum; cell spot and discocellular mark very large, with ochreous edges, the former round, filled in with Indian red, the latter oblique, broadly fabiform, with a leaden-grey centre and an Indian red inner ring; postmedial line ochreous, from three quarters of costa to near tornus, slightly dentate between veins 2 and 4, then incurved and edged with Indian red on both sides, followed by a leaden-grey patch; beyond it, a similar line, with an indian red patch proximally of it in interspace I; terminal area Indian red; subterminal line faint, whitish, undulate, with a leaden-grey patch before it, below apex, and another between veins $R_3$ and $M_1$; just before termen, a fine whitish line; cilia blackish-brown with some grey and a fine pale basal line. Hind wings orange-yellow, without discocellular spot; terminal area broadly blackish-brown, extending costally to before middle, incurved on vein 3. Length of fore wing 17 – 26 mm.

Early stages: Larva ochreous brown with strong black longitudinal lines; first and eleventh segments yellow or orange above, spotted with black; lateral tufts of blue hairs, and scattered long pale hairs on shiny black warts. On *Vitis trifolia*.

Distribution: West China to North India, Bhutan, Sikkim, Java, Borneo, Malaya, Bali.

### Sarbanissa transiens transiens (WALKER)

1912 *Seudyra transiens transiens* – JORDAN in SEITZ, Großschmetterl. Erde, 11.
1874 *Agarista aegoceroides* FELDER, Reise Novara: pl. 107, fig. 10. – Terra typica: not known (Type in British Museum).
1892 *Seudyra aegoceroides* – KIRBY, Syn. Cat. Lepidopt. Het.:38.

Diagnosis: As above.

Distribution: Malaya, Sumatra, Borneo, Java (terra typica), Bali.

### Sarbanissa transiens subalba (LEECH)

1890 *Seudyra subalba* LEECH, Entomologist, **23**:110 (♂). – Terra typica: Changyang, China (Type, ♂, in British Museum).
1912 *Seudyra transiens subalba* – JORDAN in SEITZ, Großschmetterl. Erde, **11**:27.

Diagnosis: Hind wings with the terminal band narrower on underside and not reaching the middle of costa.

Distribution: West China (South Hupé [Changyang = terra typica], Hunan, Chekiang, North Kwangtung, West Yunnan).

### Sarbanissa transiens dissimilis (SWINHOE)

1890 *Seudyra dissimilis* SWINHOE, Trans. Entomol. Soc. London, **1890**:174 (♂). – Terra typica: Mandalay, Burma (Type, ♂, in British Museum).
1912 *Seudyra transiens dissimilis* – JORDAN in SEITZ, Großschmetterl. Erde, **11**:27.

Diagnosis: Hind wings with the marginal band slightly broader than in *S. t. subalba.* Discal band on fore wings somewhat narrower on upperside.

Distribution: Burma (Mandalay = terra typica), Sikkim, Bhutan, Assam.

### Sarbanissa catocaloides (WALKER)

1862 *Phaegorista catocaloides* WALKER, J. Linn. Soc., Zool., **6**:87. – Terra typica: Sikkim (Type, ♀, in British Museum).
1892 *Seudyra transiens* – KIRBY (partim), Syn. Cat. Lepidopt. Het.:38.
1912 *Seudyra catocaloides* – JORDAN in SEITZ, Großschmetterl. Erde, **11**:28; pl. 4d.

Diagnosis: Very near *S. transiens*. Abdomen with dorsal black spots and with black dots on spiracula; female without woolly expansion at the extremity. Fore wings with two submarginal leaden-grey spots between $R_3$ and $M_2$, absent in *S. transiens*; on dorsum, very near to tornus, a spot sprinkled with leaden grey scales, preceded by a rusty-red spot cut in two by a pale line, and another leaden grey spot. Hind wings orange, with a black discocellular point; black terminal band much constricted at middle.

Distribution: Sikkim, Assam, Borneo, Nias, Java, Lombok.

### Sarbanissa melanura (JORDAN)

1912 *Seudyra melanura* JORDAN in SEITZ, Großschmetterl. Erde, **11**:28 (♂). – Terra typica: Tondano, North Celebes (Type, ♂, in British Museum).

Diagnosis: Head, palpi and thorax above brownish-black; pronotum with a fine grey edge; palpi with some greyish-white scales; fore and mid legs almost wholly black; abdomen orange, with the VI, VII and VIII urites black on underside; female with the extremity woolly expanded, creamy-yellow. Fore wings much like those in *S. transiens*, with little yellowish-grey sprinkling; discal band pale creamy-yellow, 1 mm broad; streak distally of it reaching the lower media, where it meets a triangular pale creamy-yellow spot on termen, below vein 2. Hind wings orange, without black discocellular spot; terminal band narrower than in *S. transiens* and *S. catocaloides*, regularly narrowing towards the anal margin.

Distribution: North Celebes (Tondano = terra typica).

## Genus **Zalissa** WALKER

(fig. 39)

1865 *Zalissa* WALKER, List Lepidopt. Het. Brit. Mus., **33**:936. – Species typica: *Zalissa catocalina* WALKER.

Diagnosis: Proboscis fully developed; antennae of male with short fasciculate branches; palpi upturned; the second joint reaching to middle of frons, moderately fringed with hair in front; the third joint moderate, porrect; eyes large, round; frons with a rounded prominence, without raised rims, replaced by a few rounded gibbosities, and with a corneous plate below it; thorax clothed with hair and scales mixed, without crests; tibiae fringed with long hair; abdomen with large dorsal crests on the basal segments. Fore wings with the termen slightly crenulate; nervation: vein 2 from seven-ninths of cell; 3, 4, 5 approximate, equidistant; 6 from the upper cell-angle; areole small, elongate and narrow; 7 and the very short stalk of 8 + 9 from the extremity of the areole; 10 from its upper angle. In hind wings, vein 2 from three-quarters of cell; 3, 4 and 6, 7 from the cell-angles; discocellulars incurved; 5 obsolescent, from middle of discocellulars.

Male genitalia: Uncus about four-fifths of costa in length, slender, arched, with a minute terminal hook. Tubus analis broad, weakly sclerified. Vincula rather narrow, broader proximally, with moderate bunches of stiff hair. Valva banana-shaped; costa distinctly outcurved soon after middle, but nearly straight terminally; apex forming an angle of 45°; termen and sacculus regularly outcurved; there are only about four subapical setae; fold of sacculus moderate, reaching to middle of valva; harpe about two-thirds of uncus in lenght, arched. Aedeagus one half again as long as the costa, spoon-shaped proximally; terminal portion with an elongate patch of minute cornuti; fultura inferior produced into a short process and with the proximal angles faintly produced. Saccus semi-elliptical.

Distribution: Australia, New Guinea.

### **Zalissa catocalina** WALKER

1865 *Zalissa catocalina* WALKER, List Lepidopt. Het. Brit. Mus., **33**:936 (♀). – Terra typica: Western Australia (Type, ♀, in British Museum).

Diagnosis: Head grey; thorax red-brown with a greyish dorsal fascia; pectus white; legs with the pilosity mixed white and brown; tarsi brown ringed with white; abdomen orange-yellow; crests reddish-brown; underside white banded

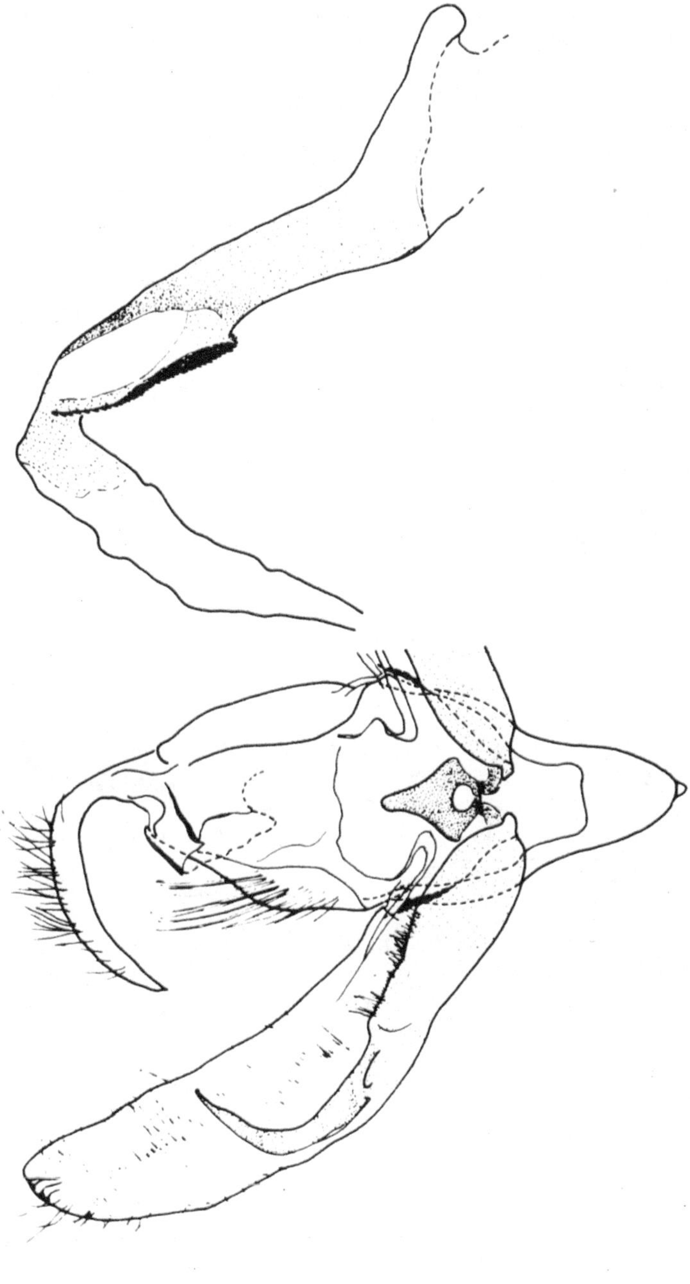

Fig. 39. *Zalissa catocalina* WALKER. Type of *Z. c. pratti* B. M. Noct. 8471

with black. Fore wings whitish, sprinkled and suffused with red-brown and dark-brown; basal area and dorsal area below the submedian fold with a purplish gloss; a curved white or brownish patch beyond the discocellular mark; terminal area whiter from apex to vein 2; subbasal line double, waved, running from costa to the submedian fold, then inwardly oblique and slightly excurved below vein 1; cell-mark with a faint pale ringlet, round; subcellular mark with a dark centre and rather diffuse white ringlet; postmedial line double, strongly bent outwards below the costa, broadly outcurved to vein 2, then slightly waved; subterminal line curved, very slightly waved, white, very finely edged on both sides with brown. Hind wings orange-yellow; terminal area blackish-brown, at costa extending to middle, narrowing to tornus, slightly incurved at cell and the postmedian fold, outcurved in between, sometimes more or less evenly incurved; some white on termen; a chocolate-red line on termen towards tornus, with a white line before it; cilia white from vein 7 to the discal fold, then blackish with a fine white line at base. Length of forewing 18 mm.

Distribution: Australia; New Guinea.

### Zalissa catocalina catocalina WALKER

1912 *Zalissa catocalina* subsp. *catocalina* – JORDAN in SEITZ, Großschmetterl. Erde, **11**:28; pl. 4d.

Diagnosis: Fore wings with the postmedial spot white; underside with a white patch between the cell and discocellular marks.

Distribution: West Australia.

### Zalissa catocalina pratti (BETHUNE–BAKER)

1906 *Hypocala pratti* BETHUNE–BAKER, Novit. Zool., **13**:249 (♂). – Terra typica: New Guinea (Type, ♂, in British Museum).

1910 *Zalissa pratti* HAMPSON, Cat. Lepidopt. Phal. Brit. Mus., **9**:442; pl. 146, fig. 13.

Diagnosis: Fore wings with the postmedial spot pale greyish-brown; underside without distinct white patch between the cell- and discocellular marks; hind wings with the blackish-brown terminal band narrower, reduced towards the tornus to sprinkling, and with its proximal edge more or less regularly incurved.

Distribution: New Guinea (northwestwards to Arfak Mountains).

# Alphabetic Register of Scientific Names

# DAS TIERREICH

## Eine Zusammenstellung und Kennzeichnung der rezenten Tierformen

Herausgeber
Robert Mertens und Willi Hennig
Schriftleiter: Heinz Wermuth

| g. | Bearbeiter | Titel | Ersch.-jahr | Umfang | Preis in DM |
|---|---|---|---|---|---|
| | A. Schmidt | COLEOPTERA: *Aphodiinae* | 1922 | VI, 614 Seiten, 34 Abbildungen | 120,– |
| | F. Nieden | AMPHIBIA: *Anura I* | 1923 | XXXII, 584 Seiten, 380 Abbildungen | 115,50 |
| | W. Kükenthal | COELENTERATA: *Gorgonaria* | 1924 | XXVIII, 478 Seiten, 209 Abbildungen | 144,– |
| | J. J. Kieffer | HYMENOPTERA: Proctotrupoidea: *Scelionidae* | 1926 | XXXVI, 885 Seiten, 340 Abbildungen | 300,– |
| | F. Nieden | AMPHIBIA: *Anura II Engystomatidae* | 1926 | XVI, 119 Seiten, 55 Abbildungen | 42,– |
| | E. Giglio-Tos | ORTHOPTERA: *Mantidae* | 1927 | XL, 707 Seiten, 6 Abbildungen | 247,50 |
| | H. Stichel | LEPIDOPTERA: *Nemeobiinae* | 1928 | XXX, 300 Seiten, 197 Abbildungen | 126,50 |
| | C. Graf Attems | MYRIAPODA: *I. Geophilomorpha* | 1929 | XXIII, 388 Seiten, 307 Abbildungen | 180,– |
| | F. Kiefer | CRUSTACEA COPEPODA: *II. Cyclopoida Gnathostoma* | 1929 | XVI, 102 Seiten, 42 Abbildungen | 56,30 |
| | C. Graf Attems | MYRIAPODA: *II. Scolopendromorpha* | 1930 | XIX, 308 Seiten, 387 Abbildungen | 150,– |
| | E. Ahl | AMPHIBIA: *Anura III Polypedatidae* | 1931 | XVI, 477 Seiten, 320 Abbildungen | 232,50 |
| | S. Thor | ARACHNOIDEA ACARINA: *Bdellidae, Nicoletiellidae, Cryptognathidae* | 1931 | XIII, 87 Seiten, 93 Abbildungen | 56,20 |
| | M. Beier | PSEUDOSCORPIONIDEA I: Subord: *Chthoniinea* et *Neobisiinea* | 1932 | XX, 258 Seiten, 271 Abbildungen | 120,– |
| | M. Beier | PSEUDOSCORPIONIDEA II: Subord. *Cheliferinea* | 1932 | XXI, 294 Seiten, 300 Abbildungen | 146,20 |

| Lfg. | Bearbeiter | Titel | Ersch.-jahr | Umfang | Preis in DM |
|---|---|---|---|---|---|
| 59 | M. Holly | CYCLOSTOMATA | 1933 | XIV, 62 Seiten, 57 Abbildungen | 37,5 |
| 60 | S. Thor | ACARINA: *Tydeidae, Ereynetidae* | 1933 | XI, 84 Seiten, 102 Abbildungen | 45,- |
| 61 | M. Holly | PISCES: *2. Dipnoi* | 1933 | IX, 20 Seiten, 25 Abbildungen | 10,- |
| 62 | F. Werner | REPTILIA: *Loricata* | 1933 | XIV, 40 Seiten, 33 Abbildungen | 26,2 |
| 63 | M. Holly | PISCES: *4. Crossopterygii* | 1933 | X, 23 Seiten, 30 Abbildungen | 10,- |
| 64 | F. Bryk | LEPIDOPTERA: *Baroniidae, Teinopalpidae, Parnassiidae I* | 1934 | XXIII, 131 Seiten, 87 Abbildungen | 75,- |
| 65 | F. Bryk | LEPIDOPTERA: *Parnassiidae II* (Subfam. *Parnassiinae*) | 1935 | LI, 790 Seiten, 698 Abbildungen | 412,5 |
| 66 | E. Marcus | ARTHROPODA: *Tardigrada* | 1936 | XVI, 329 Seiten, 306 Abbildungen | 150,- |
| 67 | M. Holly | PISCES: *4. Ganoidei* | 1936 | XVI, 65 Seiten, 56 Abbildungen | 37,5 |
| 68 | C. Graf Attems | MYRIAPODA: *3. Polydesmoidea I.* Fam. *Strongylosomidae* | 1937 | XXII, 300 Seiten, 343 Abbildungen | 150,- |
| 69 | C. Graf Attems | MYRIAPODA: *3. Polydesmoidea II.* Fam. *Leptodesmidae, Platyrhachidae, Oxydesmidae, Gomphodesmidae* | 1938<br>1938 | XXVIII,487 Seiten, 509 Abbildungen | 243,8 |
| 70 | C. Graf Attems | MYRIAPODA: *3. Polydesmoidea III* | 1940 | XXXII,568 Seiten, 719 Abbildungen | 300,- |
| 71 | S. Thor & C. Willmann | ACARINA: *3.–71 a: Eupodidae, Penthalodidae, Penthaleidae, Rhagidiidae, Pachygnathidae, Cunaxidae. – 72 b Trombidiidae* | 1941/1947 | XXXVI,541 Seiten, 599 Abbildungen | 389,2 |
| 72 | H. Wermuth | REPTILIA: *Helodermatidae* | 1958 | 16 Seiten, 9 Figuren | 16,- |
| 73 | M. Beier | ORTHOPTERA: *Tettigoniidae* (Pseudophyllinae I) | 1962 | XII, 468 Seiten, 245 Abbildungen | 210,- |
| 74 | M. Beier | ORTHOPTERA: *Tettigoniidae* (Pseudophyllinae II) | 1960 | 396 Seiten, 241 Figuren | 262,5 |
| 75 | K. Odening | TREMATODA (Digenea): *Plagiorchiidae* III (Haematoloechinae) und *Omphalometridae* | 1960 | IV, 77 Seiten, 60 Figuren | 64,- |

| fg. | Bearbeiter | Titel | Ersch.-jahr | Umfang | Preis in DM |
|---|---|---|---|---|---|
| 5 | K. Sanft | AVES (Upupae): *Bucerotidae* | 1960 | IV, 176 Seiten, 106 Figuren | 156,– |
| 7 | R. Kilias | GASTROPODA PROSOBRANCHIA: *Tonnidae* | 1962 | 63 Seiten, 47 Figuren | 28,– |
| 8 | St. W. Gorham | GYMNOPHIONA. *Liste der rezenten Amphibien und Reptilien* | 1962 | X, 25 Seiten | 15,– |
| 9 | R. Mertens | HELODERMATIDAE, VARANIDAE, LANTHANOTIDAE. *Liste der rezenten Amphibien u. Reptilien* | 1963 | X, 26 Seiten | 14,– |
| 0 | H. Wermuth | GEKKONIDAE, PYGOPODIDAE, XANTUSIIDAE. *Liste der rezenten Amphibien und Reptilien* | 1965 | XXII, 246 Seiten | 117,80 |
| 1 | J. A. Peters | COLUBRIDAE (Dipsadinae). *Liste der rezenten Amphibien und Reptilien* | 1965 | VIII, 18 Seiten | 12,50 |
| 2 | J. Illies | PLECOPTERA, *Katalog* | 1966 | XXX, 632 Seiten, 20 Abbildungen | 294,– |
| 3 | R. Mertens | CHAMAELEONIDAE. *Liste der rezenten Amphibien und Reptilien* | 1966 | X, 37 Seiten | 26,– |
| 4 | C. Gans | UROPELTIDAE. *Liste der rezenten Amphibien und Reptilien* | 1966 | 29 Seiten | 14,– |
| [illegible] | St. W. Gorham | ASCAPHIDAE, LEIOPELMATIDAE, PIPIDAE, DISCOGLOSSIDAE, PELOBATIDAE, LEPTODACTYLIDAE, RHINOPHRYNIDAE. *Liste der rezenten Amphibien und Reptilien* | 1966 | XVI, 222 Seiten | 107,– |
|  | H. Wermuth | AGAMIDAE. *Liste der rezenten Amphibien und Reptilien* | 1967 | XIV, 127 Seiten | 72,– |
|  | H. Wermuth | CORDYLIDAE (Cordylinae + Gerrhosaurinae). *Liste der rezenten Amphibien und Reptilien* | 1968 | X, 30 Seiten | 24,– |
|  | F. Haas | BIVALVIA: *Unionacea* | 1969 | X, 663 Seiten, 5 Abbildungen | 380,– |
|  | A. F. Stimson | BOIDAE, (Boinae + Bolyeriinae + Loxoceminae + Pythoninae). *Liste der rezenten Amphibien und Reptilien* | 1969 | XI, 49 Seiten | 38,– |

| Lfg. | Bearbeiter | Titel | Ersch.-jahr | Umfang | Preis in DM |
|---|---|---|---|---|---|
| 90 | H. Wermuth | ANGUIDAE, ANNIELLIDAE, XENOSAURIDAE. *Liste der rezenten Amphibien und Reptilien* | 1969 | XII, 41 Seiten | 33,- |
| 91 | M. Fischer | HYMENOPTERA: *Braconidae* (Opiinae I) | 1972 | XII, 620 Seiten, 463 Figuren | 580,- |
| 92 | R. Kilias | GASTROPODA PROSOBRANCHIA: *Cymatiidae* | 1973 | VIII, 235 Seiten, 149 Figuren | 288,- |
| 93 | P. Banarescu & T. T. Nalbant | PISCES, TELEOSTEI: *Cyprinidae (Gobioninae)* | 1973 | VII, 304 Seiten, 154 Figuren, 19 Karten | 320,- |
| 94 | P. Zwick | PLECOPTERA, *Phylogenetisches System und Katalog* | 1973 | XXXII, 465 Seiten, 75 Abbildungen | 490,- |
| 95 | E. Duellman | HYLIDAE, CENTROLENIDAE, PSEUDIDAE. *Liste der rezenten Amphibien und Reptilien* | 1976 | XIX, 225 Seiten | 280,- |
| 96 | M. Fischer | HYMENOPTERA: *Braconidae* (Opiinae II / Amerika) | 1976 | etwa 1038 Seiten, 890 Figuren | 980,- |
| 97 | S. G. Kiriakoff | LEPIDOPTERA NOCTUIFORMES: *Agaristidae I* | 1976 | IX, 180 Seiten, 39 Figuren | 150,- |
| 98 | S. G. Kiriakoff | LEPIDOPTERA NOCTUIFORMES: *Agaristidae II* | 1976 | VIII, 165 Seiten, 52 Figuren | 140, |
| 99 | S. G. Kiriakoff | LEPIDOPTERA NOCTUIFORMES: *Agaristidae III* | 1976 | VI, 86 Seiten, 26 Figuren | 90, |
| 100 | H. Wermuth & † R. Mertens | TESTUDINES, CROCODYLIA, RHYNCHOCEPHALIA. *Liste der rezenten Amphibien und Reptilien* | 1976 | XXVII, 174 Seiten | 210, |

zoologique en question sans pour cela déborder le cadré de l'ensemble de l'ouvrage. Sont invités à collaborer, tous les auteurs qui désireraient publier des révisions systématiques de groupes zoologiques récents. Conformément aux voeux des auteurs, leurs traités seront imprimés en anglais, en français et en allemand. Les honoraires des auteurs sont fixés à DM 55,— par placard de l'imprimerie (= 16 pages) et 5 exemplaires gratuits après parution de leurs traités. Pour plus amples détails, se reporter à les livraisons No. 72 et suivantes qui vient de paraître depuis 1958. Les éditeurs fournissent tous renseignements sur demande ou des indications précises à l'intention des auteurs. Dans la série „DAS TIERREICH", on publiera aussi des révisions sommaires de groupes zoologiques comme „Checklists".

---

Das „TIERREICH" wurde von der Deutschen Zoologischen Gesellschaft in der Absicht gegründet, eine vollständige systematische Darstellung aller rezenten Gruppen des Tierreichs zu schaffen. Nach dem zweiten Weltkrieg wurde das Unternehmen abgebrochen, da das Ziel der Vollständigkeit unerreichbar schien. Damit entfielen jedoch Anreiz und Möglichkeit, umfassende Revisionen größerer Tiergruppen in fortlaufenden Lieferungen zu veröffentlichen. Aus diesem Grunde hat sich der Verlag Walter de Gruyter & Co. (Berlin) entschlossen, das „TIERREICH" zwar grundsätzlich unter den ursprünglichen Gesichtspunkten, darüber hinaus aber lediglich als eine zwanglose Reihenfolge systematischer Monographien wieder aufleben zu lassen, ohne zu beabsichtigen, alle zoologischen Gruppen nach einem feststehenden Plan abzuhandeln. Diese Monographien sollen sich nicht ausschließlich auf die Belange der Taxionomie beschränken, sondern auch anatomische, biologische, physiologische und andere Tatsachen berücksichtigen, sofern sie geeignet sind, die betreffende Tiergruppe näher zu charakterisieren, ohne den Rahmen des Gesamtwerkes zu sprengen. Zur Mitarbeit sind alle Autoren eingeladen, die den Wunsch haben, systematische Revisionen rezenter Tiergruppen zu veröffentlichen. Entsprechend den Wünschen der Verfasser werden ihre Abhandlungen in Englisch, Französisch und Deutsch gedruckt. Die Autoren erhalten ein Honorar in Höhe von DM 55,— für jeden Druckbogen (= 16 Seiten) und 5 Sonderdrucke nach Erscheinen ihrer Beiträge. Weitere Einzelheiten mögen den seit 1958 erschienenen Lieferungen Nr. 72 und folgenden entnommen werden. Auf Wunsch erteilen die Herausgeber jegliche Auskünfte oder geben genaue Anweisungen an die Autoren ab. In der Reihe „DAS TIERREICH" erscheinen auch summarische Bearbeitungen von größeren Tiergruppen in Form sogenannter „Checklists".

Die Herausgeber

*†Prof. Dr. Robert Mertens (Wirbeltiere)*
Natur-Museum und Forschungs-Institut Senckenberg
Senckenberg-Anlage 25, 6000 Frankfurt am Main 1

*Prof. Dr. Willi Hennig (Wirbellose)*
Staatliches Museum für Naturkunde
Arsenalplatz 3, 7140 Ludwigsburg

*Dr. Heinz Wermuth (Schriftleiter)*
Staatliches Museum für Naturkunde
Arsenalplatz 3, 7140 Ludwigsburg

www.ingramcontent.com/pod-product-compliance
Lightning Source LLC
LaVergne TN
LVHW042241300926
849161LV00003B/432